Ohio Covered Bridges

by

Harold Stiver

Copyright Statement

Ohio Covered Bridges

A Guide for Photographers and Explorers

Version 1.0

ISBN # 978-1-927835-03-6

Table of Contents

How to use this Book

For each of the 145 historical or Traditional Covered Bridges remaining in Ohio, we have included photographs as well as descriptive and statistical data. Traditional Covered Bridges are those that follow the building practices of the Nineteenth Century and the early part of the Twentieth Century or those built later that follow those methods. All of these bridges have had repairs done as portions wear out, and some may have been almost entirely replaced through the years. I have used "The National Society for the Preservation of Covered Bridges, Inc." list of what they consider as Traditional Bridges.

Following is data included for each bridge

Name: This is listed in bold type, and where there are other names, it is the common name or the name listed on an accompanying plaque.

Other Names: Underneath the Common Name in brackets, you will find other names that the bridge has been known by.

Nearest County and Township are listed.

It is frustrating to go on an excursion to see something and not be able to find it. This book offers you multiple ways to ensure that doesn't happen.

GPS Position: This is our recommended method. Enter the coordinates in a good GPS unit and it should take you right there. You, of course, must use care that you are not led off road or on a dangerous route.

Detailed Driving Directions: Directions from a town near to the bridge.

Builder: If known, the name of the original builder(s) is listed.

Year Built: As well as the year built, if it has been moved it will shown with the year preceded by the letter M and, if a major repair has been done, the year will be shown preceded by the letter R.

Truss Type: The type for the particular bridge will be listed. If you are interested in more information on the various types of trusses, access "Truss Types" from the Table of Contents.

Dimensions: The length and number of spans

Notes: A place where you can find additional items of interest about the bridge.

World Index Number:
Covered bridges are assigned a number to keep track of them which consists of three numbers separated by hyphens.

The first number represents the number of the U.S. State in alphabetical order. Following number 50 for the 50th state are additional numbers for Canadian provinces. Thus the numbers 05 represents California.

The second set of numbers represents the county of that state, again based on alphabetical order. Humboldt is the 12th county alphabetically in California, and it is designated as 05-12.

Each bridge in that county is given a number as it was discovered or built. Zane's Ranch was the fifth bridge discovered or built in the County of Humboldt, California and it therefore has the designation of 05-12-05. Sometimes you will

see the first set of numbers replaced by the abbreviation for the state, thus CA-12-05.
A bridge is sometimes substantially rebuilt or replaced and it then has the suffix #2 added to it.

National Register of Historic Places: If the bridge has registered, the date is given.

Photographing Covered Bridges

Some standard positions
Portal: Taken to show the ends of bridge or bridge opening. This view, usually symmetrical, will include various signs posted. This is also a good way to get run over, so be careful!
3/4 view: Shows both the front and sides of the bridge, and is often the most attractive.
Side view: Taken from a bank or from the river, this gives not only a nice view of the bridge but usually allows for some interesting foreground elements.
Interior view: An image taken from the interior of the bridge will show some interesting structure but there is not a lot of available light. A tripod is important and HDR processing is helpful.
Landscape View: With the bridge smaller in the frame, you can introduce the habitat around it, particularly effective with colorful autumn foliage.

Using HDR(High Dynamic Range)
HDR is a process where multiple images of varying exposure are combined to make one image.

It has a bad name with some people because many HDR images are super-saturated, a kind of digital age version of an Elvis painted on velvet. However, the process is actually about getting a full range of exposure with no burnt out highlights or blocked shadows. This is an ideal processing solution for photographing Covered Bridges where you often have open light sky set against dark shadowed landscape and structure.

I use a series of three exposures at levels of -1 2/3, 0, +1 2/3, and this normally runs the full exposure range encountered. It is important to use a stable tripod.

One situation where you may need a larger series is shooting from within a bridge and using the window to frame an outside scene. The dynamic range is huge and you will need to have a series with a much larger range.

There are a number of software programs you can use to combine these images including newer editions of Photoshop. I use Photomatix which I have found very versatile and easy to use.

Best times for photographing bridges
Mornings and evenings are generally the best times for outdoor photography but the use of HDR processing makes it easier even in bright direct light. Although any season is good for bridge photography including the winter, fall foliage included in a scene can be spectacular.

A Short History of Covered Bridges

Let's deal with that often posed question; "Why were the bridges covered"

1. Crossing animals thought it was a barn and entered easily. I like this suggestion, it shows imagination. However, its not the answer although the original bridges normally had no windows and this is said to be because animals would not be spooked by the sight of the water.

2. To cover up the unsightly truss structure. I don't think those early pioneers were that sensitive, and personally, I like the look of the trusses.

3. To keep snow off the travelled portion. In fact the bridge owners often paid to have the insides "snowed" in order to facilitate sleighs.

4. It offered some privacy to courting couples, hence "kissing bridges". That is a nice romantic notion but no.

In fact, the bridge was covered for economic reasons. The truss system was where much of the bridge's cost was found, and if left open to the elements, it deteriorated and the bridge became unstable and unsafe. Covering it protected this valuable portion and the roof could be replaced as needed with inexpensive materials and unskilled labour. Without coverings, a bridge might only have a life span of a decade while one that was covered often lasted 75 years or more before repairs became necessary. Besides extending the longevity of a bridge, wooden covered bridges had the virtue that they could be constructed of local materials and there were many available workers skilled in working with wood.

The first known Covered Bridge in North America was built in 1804 by Theodore Burr. It was called the Waterford bridge and it spanned the Hudson River in New York.
For the rest of the century and into the 20th Century, Covered bridge building boomed as the country became populated and people needed to travel between communities. The cost of constructing and maintaining a bridge was normally borne by the nearby community and many bridges charged a toll as a method of offsetting these costs.
The period from 1825 to 1875 was the heyday of bridge building but near the end of that period iron bridges began to supplant them.

The number of Covered bridges may have numbered 10,000 but have now dropped to about 840 spread throughout North America. Many have Historical Designations which provides them protection and many communities are interested in protecting their local historical bridges.

The Bridge Builders of Ohio

If you look at the builders names in or individual bridge accounts, you will see the same names re-occurring frequently. This is because Ohio was fortunate to have a number of individuals who worked for many years building covered bridges in Ohio.

Reuben L. Partridge, born on September 10, 1823, in Wilmington, New York, came to Marysville, Ohio in 1836 where he became wagon and carriage making and later began a carpentry and furniture business. By 1855 he began building bridges and patented his own wood truss design in 1872, which was used in many of his covered bridges. he built over 125 bridges in Ohio, most of them covered and most in Union County. He died on July 17, 1900, after being injured while supervising repairs to the Alsaugh Covered Bridge and is buried at Oakdale Cemetery, Marysville, Ohio. He has the following five bridges that survive:

Bergstresser Covered Bridge
Bigelow Covered Bridge
Culberson Covered Bridge
Upper Darby CB
Spain Covered Bridge

Everett S. Sherman (1831-1897) was a covered bridge builder in Ohio. He lived and built bridges in Delaware County then moved to Preble County after a storm destroyed many of its bridges.
He used the Childs truss bridges, of which only a few examples survive in the United States. The Childs truss was patented by Horace Childs in 1846. It never became popular, but, after its patent expired, Sherman took it up for bridges in Ohio in the later part of the 19th century. He has seven bridges which survive in Ohio:

Brubaker Covered Bridge   Chambers Road Covered Bridge
Christman Covered Bridge Geeting Covered Bridge
Harshman Covered Bridge Warnke Covered Bridge
Dixon Branch Bridge

Robert W. Smith was born in Ohio in 1833 worked as a timber framer in his early years and in 1867 he formed the Smith Bridge Company and located it in Toledo, Ohio.

He built most of his bridges in Ohio and Indiana and patented designs for the Trusses of them. He would cut and fit the bridge members in his yard in Toledo and then disassemble them and ship them to the bridge site by train or boat. They would then be assembled by his crew or local carpenters under his company's direction.

The Smith Bridge Company built a large number of covered bridges in Ohio of which the following 8 remain:

Brown Covered Bridge	Buckeye Furnace CB
Engle Mill Covered Bridge	Feedwire Road CB
Jaspar Road Covered Bridge	Otway Covered Bridge
Rinard Covered Bridge	Stevenson Road CB

Ohio County Map

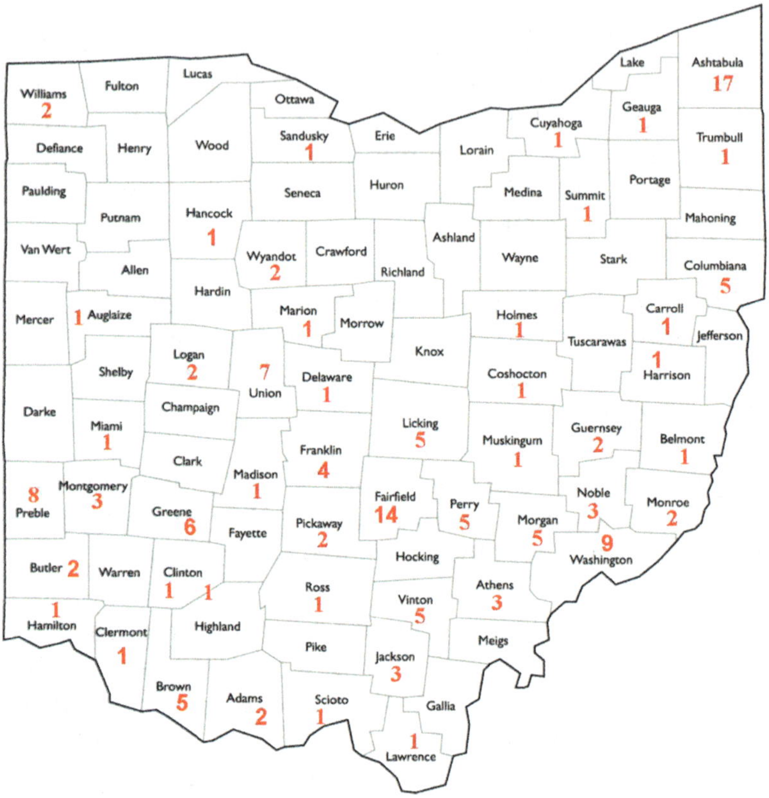

Harshaville Covered Bridge
Township: Oliver
County: Adams

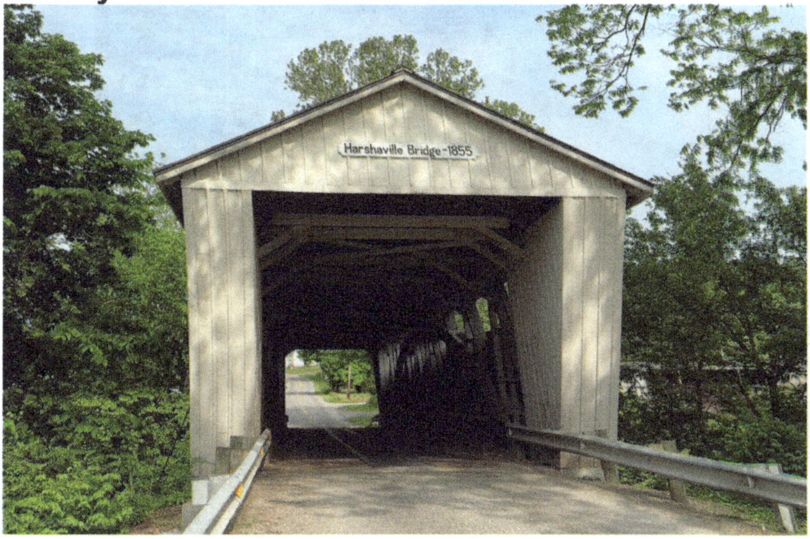

GPS Position: N 38° 54.481' W 83° 32.672'
Directions: From Seamen, go south on OH-247 for 1.2 mi and turn left on CR-1/Graces Run Rd. And the bridge is 1.3 mi
Crosses: Ohio Brush Creek, Cherry Fork
Carries: Graces Run Rd.
Builder: Not known
Year Built: CA.1855 (R1960) (R2013)
Truss Type: Multiple Kingpost and Burr Arch
Dimensions: 1 plus Span, 110 feet
Notes: The structure survived a Confederate raid in 1863 by Morgan's Raiders which caused the burning of many of the area bridges as well as the Harshaville General Store. The 2013 renovation put in concrete abutments, steel I-beams as well as replacing much of the roof and sides.

World Index Number: OH/35-01-02
National Register of Historic Places: March 16, 1976

Governor Kirker Covered Bridge

Township: Liberty
County: Adams

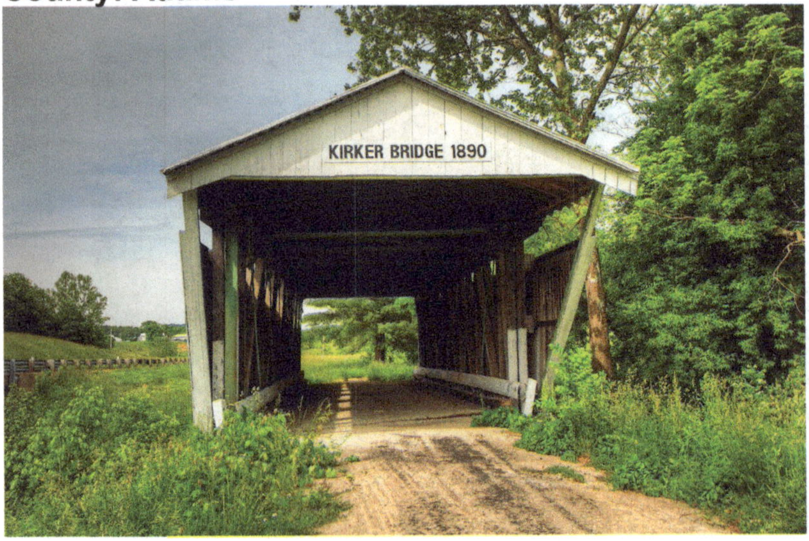

GPS Position: N 38° 47.022' W 83° 36.218'
Directions: From West Union, go SW on OH-41 for 3.0 miles and turn right on OH-136 where the bridge is 1.6 mi
Crosses: Eagle Creek, East Fork
Carries: OH-136 (Bypassed Section)
Builder: Not known
Year Built: 1890 (R1950)
Truss Type: Multiple Kingpost
Dimensions: 1 Span, 63 feet

Notes: The bridge is names for Governor Kirker who served as Ohio Governor from 1807-1808. The bridge is showing signs of wear. It is no longer in use, having been bypassed by a modern bridge beside it.

World Index Number: OH/35-01-10
National Register of Historic Places: October 19, 1975

Olins (Dewey Road) Covered Bridge
Township: Plymouth
County: Ashtabula

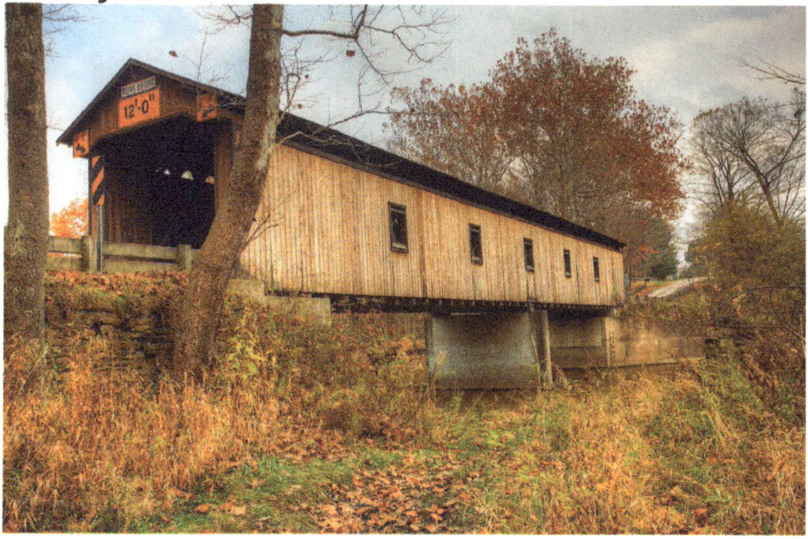

GPS Position: N 41° 51.782' W 80° 43.293'
Directions: From Kingsville go west on OH-84/Ridge Rd for 1.7 mi and turn left on Stevens Rd, then go 0.3 mi and turn right to stay on Stevens Rd which becomes Hadlock Rd. In 1.5 mi turn left on Dewey Rd where you will find the bridge.
Crosses: Ashtabula River
Carries: Dewey Road
Builder: Not known
Year Built: 1873 (R1981) (R1985) (R1992)
Truss Type: Town
Dimensions: 1 Plus Span, 115 feet
Notes: Some steel supports and a concrete center pier have been added during renovations. The five nicely balanced windows set off the unpainted sides well.

World Index Number: OH/35-04-03
National Register of Historic Places: Not listed

Creek Road Covered Bridge

County: Conneaut
Township: Ashtabula

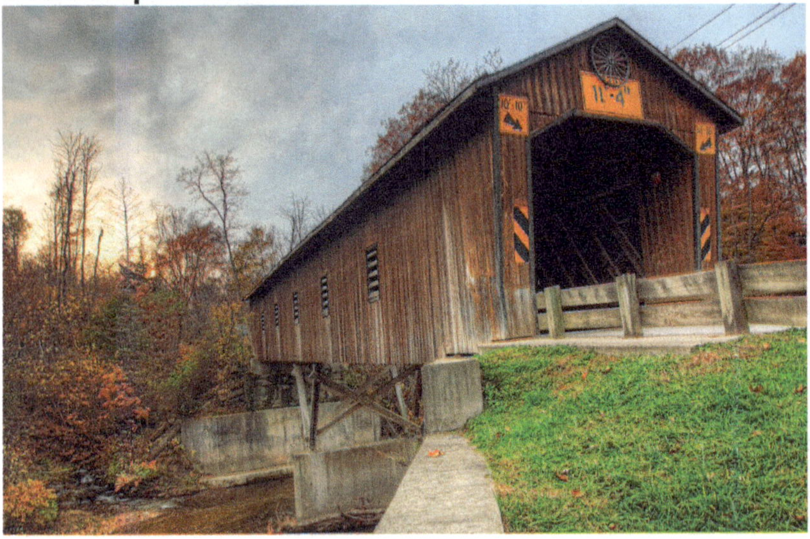

GPS Position: N 41° 55.198' W 80° 36.548'
Directions: From Kingsville go NW on Center St, continuing on Main Rd. and after 3.6 miles, turn right on Amboy Rd. In 0.4 miles turn left on Bridgeview Ln. and then in 0.3 mi turn left on Creek Road where you will see the bridge.
Crosses: Conneaut Creek
Carries: Creek Road
Builder: Not known
Year Built: CA. 1880 (R1995)
Truss Type: Town
Dimensions: 1 Plus Span, 112 feet
Notes: A concrete pier has been added that diverts ice flow as well as distributing the weight. The five balanced windows are a nice detail. It still carries vehicle traffic.

World Index Number: OH/35-04-05
National Register of Historic Places: Not listed

Middle Road Covered Bridge
Township: Conneaut
County: Ashtabula

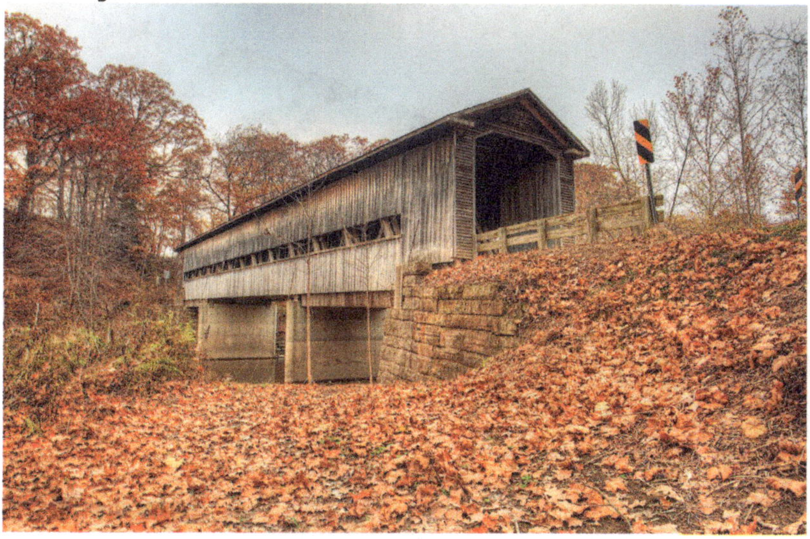

GPS Position: N 41° 54.098' W 80° 32.782'
Directions: From Farnham go east on Ridge Rd for 1.2 miles and turn right on Middle Road where you will see the bridge.
Crosses: Conneaut Creek
Carries: Middle Road
Builder: Not known
Year Built: 1868 (R1984)
Truss Type: Howe
Dimensions: 1 plus Span, 136 feet

Notes: After one end of the bridge dropped 18 inches in 1984, it was closed for repairs. It had two concrete piers added and after some timber replacement it was jacked back up. It is unpainted with a bridge length window on each side.

World Index Number: OH/35-04-06
National Register of Historic Places: Not listed

Root Road Covered Bridge
Township: Monroe
County: Ashtabula

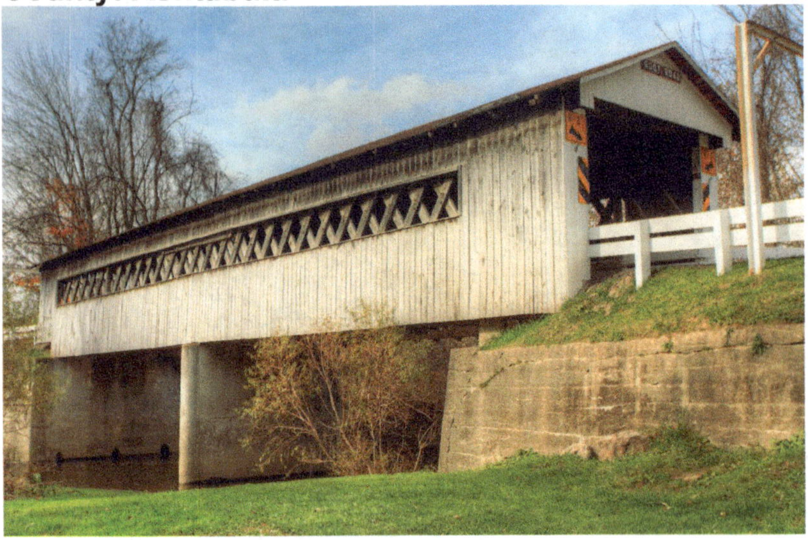

GPS Position: N 41° 50.001' W 80° 37.169'
Directions: From Kelloggsville, head south on Stanhope-Kelloggsville Rd for 1.8 mi and turn left onto Root Rd and the bridge
Crosses: Ashtabula River, West Branch
Carries: Root Road
Builder: Not known
Year Built: 1868 (R1983)
Truss Type: Town
Dimensions: 1 Plus Span, 97 feet

Notes: In 1983, the bridge was raised 18 inches and a cement center pier was added. The side windows run the length of the bridge revealing the Town trusses.

World Index Number: OH/35-04-09
National Register of Historic Places: Not listed

Benetka Road Covered Bridge

Township: Sheffield
County: Ashtabula

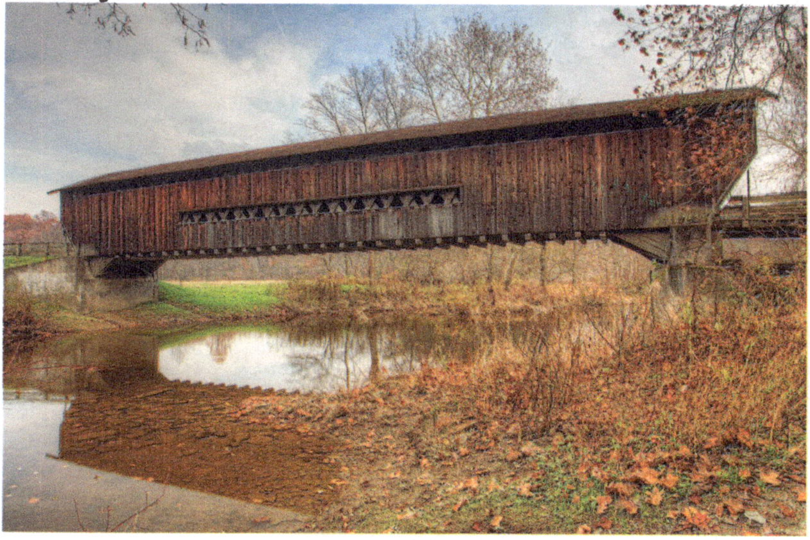

GPS Position: N 41° 50.940' W 80° 41.364'
Directions: From Kingsville go southeast on OH-193/Lake St for 2.3 mi and turn right on Plymouth Ridge Rd. Go 1.3 mi and turn left on Benetka Rd where you will find the bridge.
Crosses: Ashtabula River
Carries: Benetka Road
Builder: Not known
Year Built: Ca.1900 (R1985)
Truss Type: Town and Burr Arch
Dimensions: 1 Span, 115 feet

Notes: The 1985 renovation brought the bridge up in vehicle allowance by adding a Burr Arch and increasing the traffic height clearance. It is unpainted and looks great.

World Index Number: OH/35-04-12
National Register of Historic Places: Not listed

Graham Road Covered Bridge
Township: Pierpont
County: Ashtabula

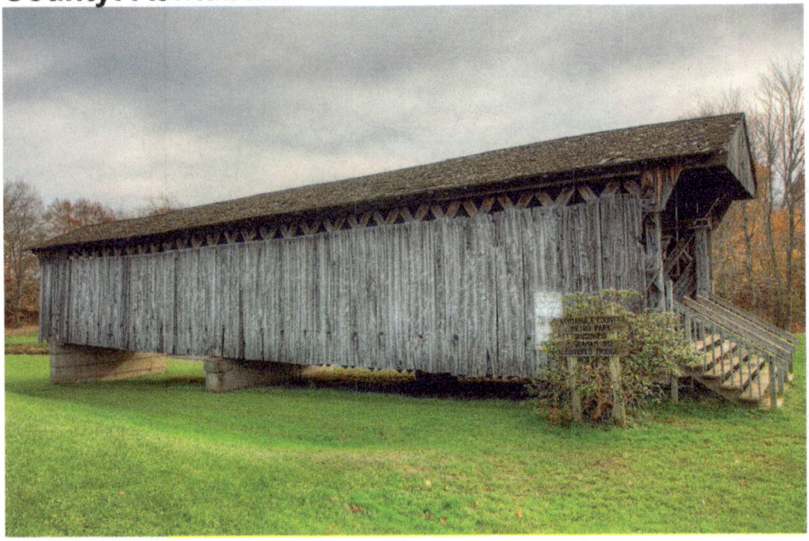

GPS Position: N 41° 46.903' W 80° 37.099'
Directions: From Jefferson go east on OH-167/Beech St for 6.8 mi and turn left on Stanhope-Kelloggsville Rd. After 2 mi turn right on Graham Rd where the bridge is a short distance.
Crosses: None
Carries: Graham Road (Bypassed)
Builder: Not known
Year Built: 1867 (R1913) (R1960s) (M1971)
Truss Type: Town
Dimensions: 1 Span, 85 feet

Notes: The bridge originally spanned the Ashtabula River in Pierpont but was moved to its present location in 1972. It was badly damaged in a flood in 1913 but repaired.

World Index Number: OH/35-04-13
National Register of Historic Places: Not listed

South Denmark Road Covered Bridge
Township: Denmark
County: Ashtabula

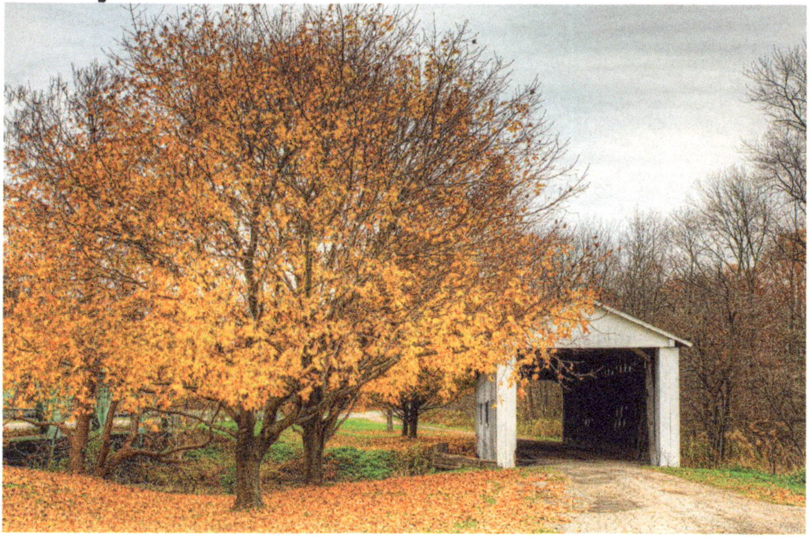

GPS Position: N 41° 43.046' W 80° 41.451'
Directions: From the town of Jefferson go east on the Jefferson St. Ext. for 0.7 miles and make a slight right onto South Denmark Rd. The bridge is 3 miles along the road.
Crosses: Mill Creek
Carries: South Denmark Rd. (Bypassed)

Builder: Not known
Year Built: 1868
Truss Type: Town
Dimensions: 1 Span, 80 feet

Notes: The bridge was bypassed in 1975 and one of the reasons was it was being damaged by trucks using it.

World Index Number: OH/35-04-14
National Register of Historic Places: Not listed

Doyle Road Covered Bridge
Township: Jefferson
County: Ashtabula

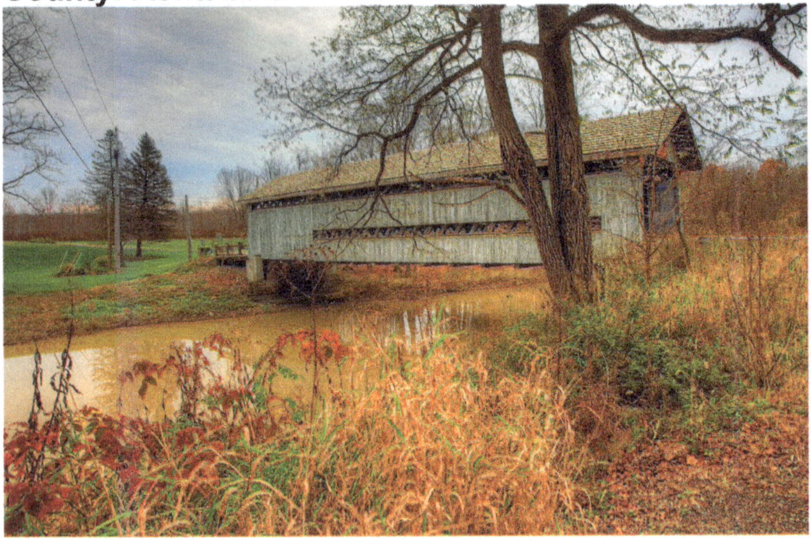

GPS Position: N 41° 45.744' W 80° 47.423'
Directions: From the town of Jefferson go west on OH-307/Beech St. for 0.6 miles and turn right on Doyle Rd. where you will find the bridge in 1 mile.
Crosses: Mill Creek
Carries: Doyle Rd.
Builder: Not known
Year Built: 1876 (R1988)
Truss Type: Town and Burr Arch
Dimensions: 1 Span, 84 feet

Notes: The laminated Burr Arch was added in the 1988 renovation as well as a new deck, roof and sides. It is open for traffic.

World Index Number: OH/35-04-16
National Register of Historic Places: Not listed

Mechanicsville Road Covered Bridge

Township: Austinburg
County: Ashtabula

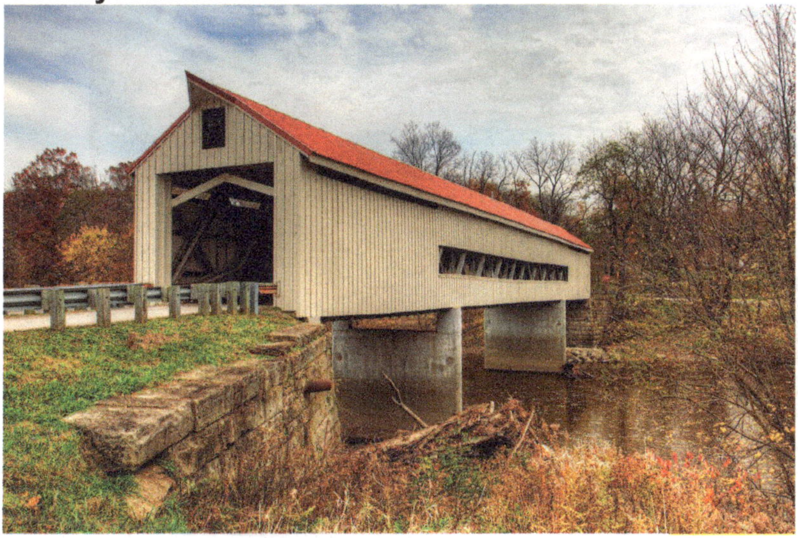

GPS Position: N 41° 45.295' W 80° 53.850'
Directions: From Austinburg go west on OH-307 for 1.8 miles and turn left on Mechanicsville Rd./Sexton Rd where you will find the bridge on a bypassed section after 1 mile.
Crosses: Grand River
Carries: Mechanicsville Rd. (Bypassed)
Builder: Not known
Year Built: 1867 (2003)
Truss Type: Howe and Burr Arch
Dimensions: 1 plus Span, 154 feet

Notes: Two concrete piers were added in 2003. It is the oldest and longest of the many covered bridges surviving in Ashtabula County

World Index Number: OH/35-04-18
National Register of Historic Places: Not listed

Harpersfield Covered Bridge
Township: Harpersfield
County: Ashtabula

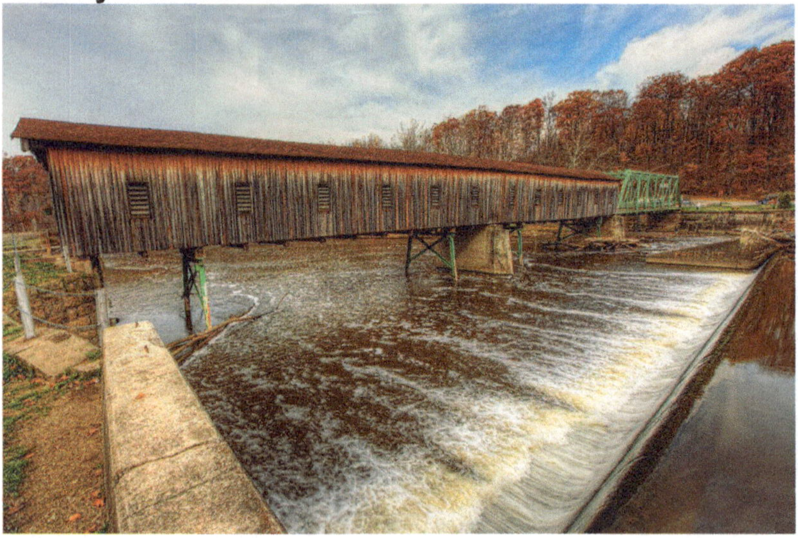

GPS Position: N 41° 45.330' W 80° 56.675'
Directions: From Geneva go south on OH-534 for 2.7 mi and turn right on State Rd./Old Hwy 534 where the bridge is a short distance.
Crosses: Grand River
Carries: State Rd./Old Hwy 534
Builder: Riverside Bridge Co
Year Built: 1868 (R1992)
Truss Type: Howe
Dimensions: 2 Plus Span, 234 feet
Notes: The pedestrian walkway was added during the 1992 renovation. Five steel piers have been added to the original two spans. It still carries a lot of traffic although a new concrete bridge has been built upstream.

World Index Number: OH/35-04-19
National Register of Historic Places: November 3, 1975

Riverdale Road Covered Bridge
Township: Morgan
County: Ashtabula

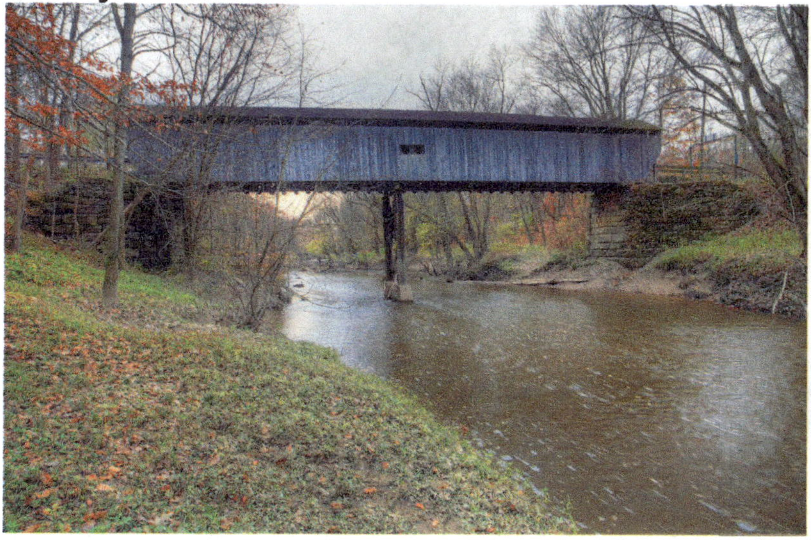

GPS Position: N 41° 40.397' W 80° 52.339'
Directions: From Rock Creek go north on OH-45/Main St for 0.6 mi and turn left on Riverdale Rd. where you find the bridge in 0.9 miles.
Crosses: Grand River
Carries: Riverdale Rd.
Builder: Not known
Year Built: 1874 (R1945) (R1965) (R1981) (R1987)
Truss Type: Town
Dimensions: 1 Plus Span, 120 feet
Notes: The bridge has had a lot of work done through the years including a steel center pier in 1945, laminated girders in 1981 and a new concrete abutment in 1981 after the east end of the bridge was washed out.

World Index Number: OH/35-04-22
National Register of Historic Places: Not listed

Warner Hollow (Windsor Mills, Wiswell Road) CB

Township: Windsor
County: Ashtabula

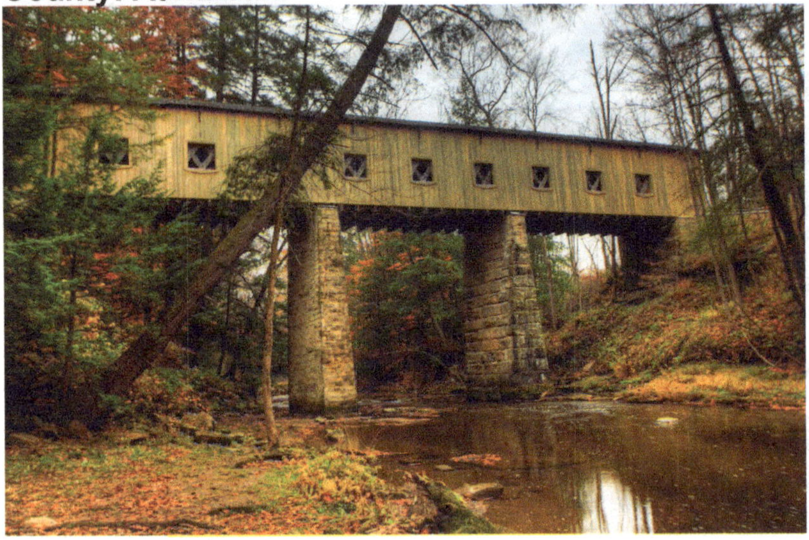

GPS Position: N 41° 32.002' W 80° 57.820'
Directions: From the town of Orwell go west on US-322/Main St for 4.7 miles and turn left on Warner Hollow Road where the bridge is a short distance.
Crosses: Phelps Creek
Carries: Warner Hollow Road
Builder: Not known
Year Built: 1867 (R2004)
Truss Type: Town
Dimensions: 3 Span, 120 feet
Notes: This is one of Ohio's finest looking bridges and the Ochre paint sets it apart. The stone piers are impressive, especially from below. After a complete renovation in 2004, it re-opened for light traffic.

World Index Number: OH/35-04-25
National Register of Historic Places: April 11, 1973

State Road Covered Bridge
Township: Monroe
County: Ashtabula

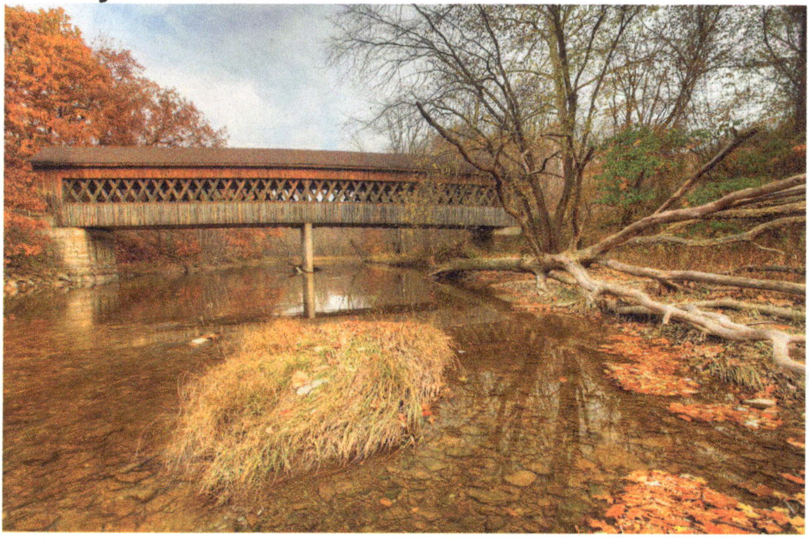

GPS Position: N 41° 53.163' W 80° 37.212'
Directions: From the town of Kingsville go east on Ridge Rd. for 2.3 miles and turn right on South Ridge Rd and then, after another 0.2 miles, right on State Road. The bridge is about 0.8 miles.
Crosses: Conneaut Creek
Carries: State Road
Builder: John Smolen, Jr
Year Built: 1983
Truss Type: Town
Dimensions: 2 Span, 157 feet
Notes: The celebrations surrounding the opening of this new bridge spurred the start of the annual Ashtabula County Covered Bridge festival.

World Index Number: OH/35-04-58
National Register of Historic Places: Not listed

Caine Road Covered Bridge
Township: Pierpont
County: Ashtabula

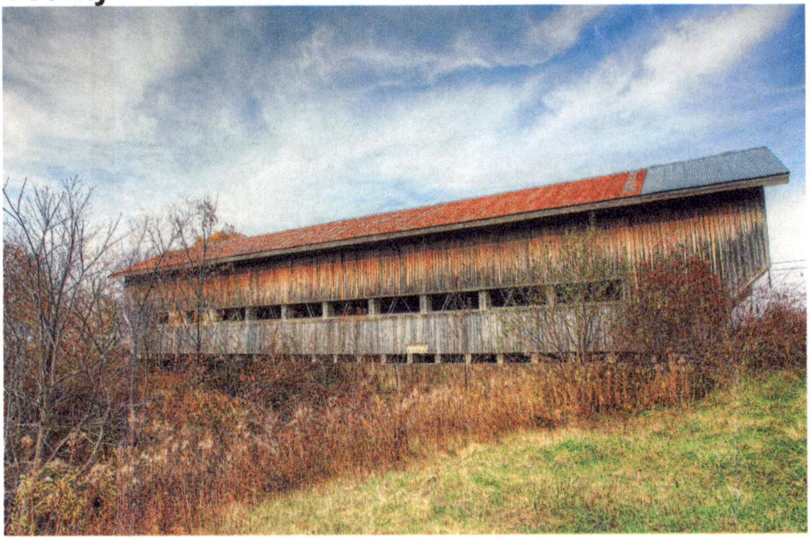

GPS Position: N 41° 45.888' W 80° 37.016'
Directions: From Jefferson go east on OH-167/Beech St for 7 mi and turn left on Stanhope-Kellogsville Rd. After 0.9 mi turn right on Caine Road where the bridge is a short distance.
Crosses: Ashtabula River, West Branch
Carries: Caine Road
Builder: Ohio Department of Transportation
Year Built: 1986
Truss Type: Pratt
Dimensions: 1 Span, 96 feet
Notes: The full length extended window sets off the bridge nicely. This is one of four new bridges that have been built in Ashtabula County. It was built in honor of Ashtabula County's 175th anniversary.

World Index Number: OH/35-04-61
National Register of Historic Places: Not listed

Giddings Road Covered Bridge

Township: Jefferson
County: Ashtabula

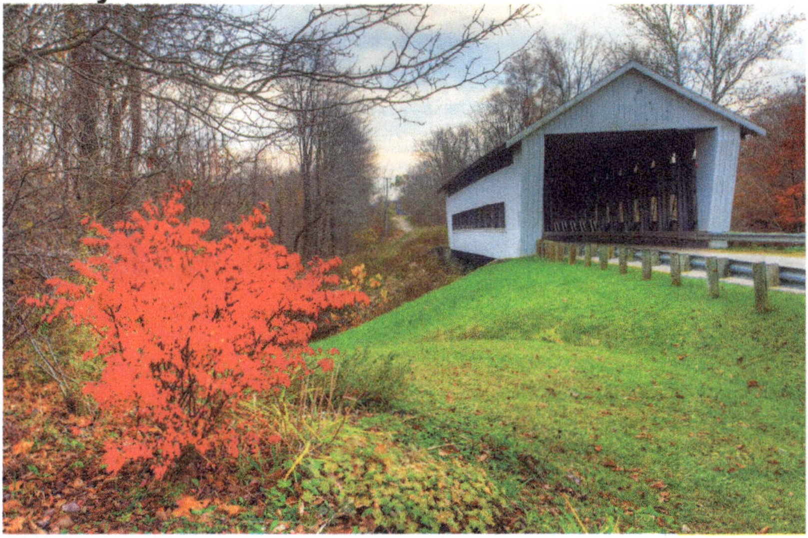

GPS Position: N 41° 45.928' W 80° 43.951'
Directions: From Jefferson go east on OH-167/Beech St for 1.1 miles and turn left on Beech St/Jones Rd. After 1.4 mi turn left on Giddings Rd. and the bridge in 0.4 miles
Crosses: Mill Creek
Carries: Giddings Rd.

Builder: Ohio Department of Transportation
Year Built: 1995
Truss Type: Pratt
Dimensions: 1 Span, 107 feet
Notes: This is one of four new covered bridges that have been built in Ashtabula County. The long row of side windows is attractive and the strong Pratt truss system is interesting.

World Index Number: OH/35-04-62
National Register of Historic Places: Not listed

Netcher Road Covered Bridge

Township: Jefferson
County: Ashtabula

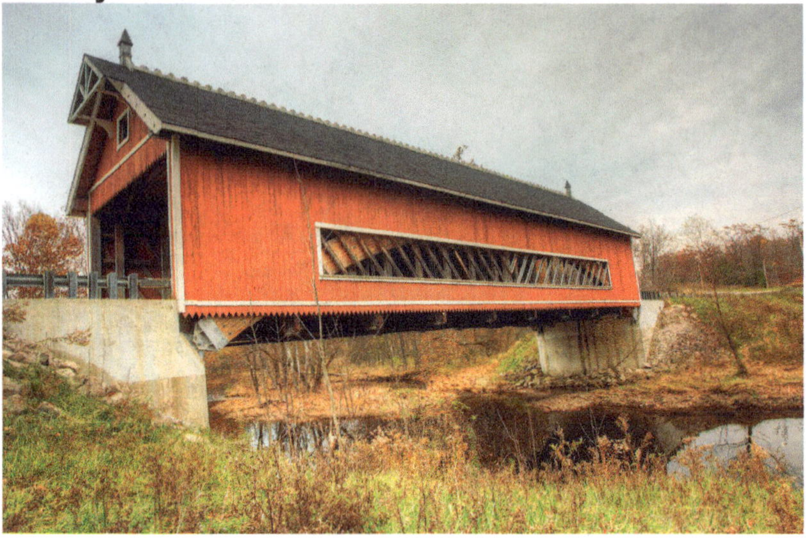

GPS Position: N 41° 44.235' W 80° 43.935'
Directions: From Jefferson go east on OH-167/Beech St for 1.3 mi and turn right on Garrett Rd. Follow Garrett Rd for 0.9 mi and turn left on Denmark Rd., then left on Netcher Rd to find the bridge.
Crosses: Mill Creek
Carries: Netcher Rd.
Builder: John Smolen Jr.
Year Built: 1999
Truss Type: Haupt Truss and Hinged Arch
Dimensions: 1 Span, 104 feet
Notes: This structure has a long laminated arch which has a hinge at the top which joins two pieces. One of four new covered bridges built in Ashtabula County.

World Index Number: OH/35-04-63
National Register of Historic Places: Not listed

Smolen-Gulf Covered Bridge
Township: Ashtabula-Plymouth
County: Ashtabula

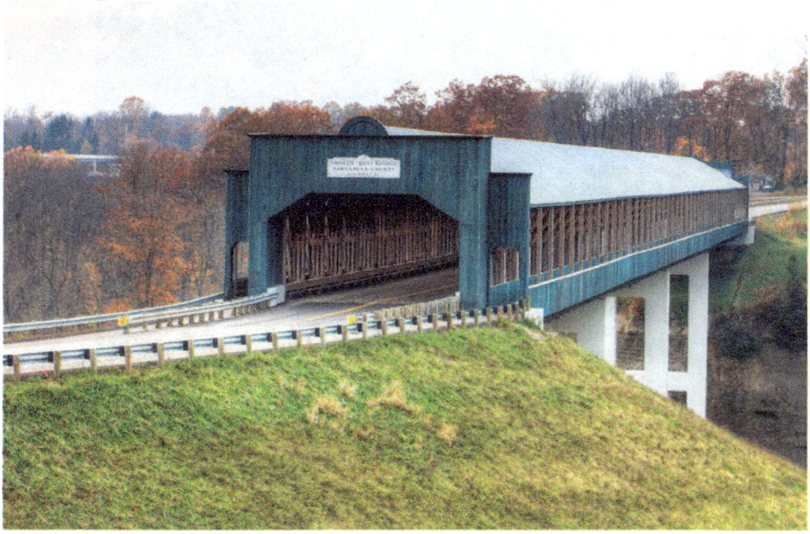

GPS Position: N 41° 51.376' W 80° 45.574'
Directions: From the town of Ashtabula go east on State Road and you will find the bridge in 0.4 miles
Crosses: Ashtabula River
Carries: State Road
Builder: Union Industrial Contractors and Koski Construction
Year Built: 2008
Truss Type: Pratt
Dimensions: 4 Span, 613 feet
Notes: A massive bridge and the longest in the U.S., four spans and 613 feet in length with a Pratt truss system. It accommodates two lanes of traffic and 2 pedestrian walkways. It carries interstate traffic loads and cost about 8 million dollars to build.

World Index Number: OH/35-04-64
National Register of Historic Places: Not listed

Palos (Newton) Covered Bridge
Township: Trimble
County: Athens

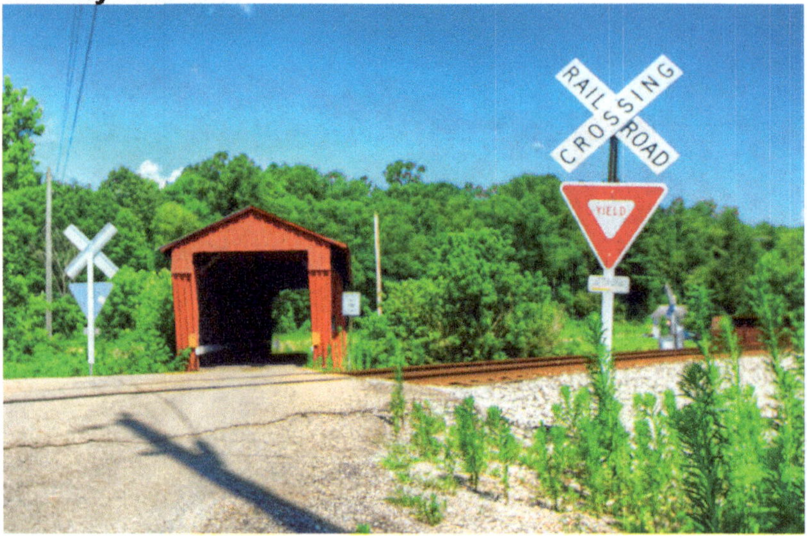

GPS Position: N 39° 31.520' W 82° 04.300'
Directions: From Glouster, head north on OH-13 N for 1.3 mi and turn right on Red Rock Rd and you will see the bridge
Crosses: Sunday Creek
Carries: Red Rock Rd

Builder: Daniel Fulton
Year Built: 1875 (R2003)
Truss Type: Multiple Kingpost
Dimensions: 1 Plus Span, 75 feet

Notes: It was built to access the W. P. Rice Mine, a coal mine which operated from 1913-1925

World Index Number: OH/35-05-01
National Register of Historic Places: November 11, 1977

Kidwell (Monserat) Covered Bridge

Township: Dover
County: Athens

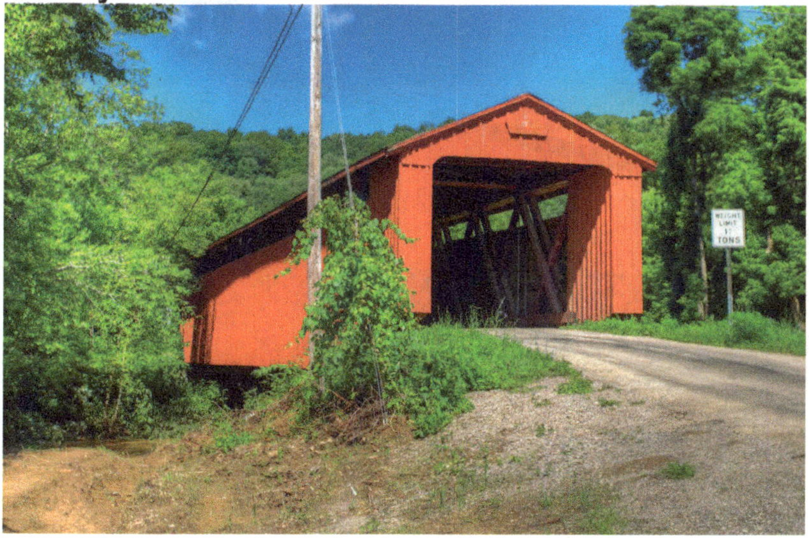

GPS Position: N 39° 27.340' W 82° 06.161'
Directions: From the town of Millfield, go north on CR-93 for 0.6 miles and turn left on OH-13. After 1.0 miles turn left on Monserat Road where you will find the bridge in 0.2 miles.
Crosses: Sunday Creek
Carries: Monserat Road
Builder: August Borneman and Hocking Valley Bridge Works
Year Built: 1880 (R2003)
Truss Type: Howe
Dimensions: 1 Span, 96 feet
Notes: The bridge was closed in 1978 after being damaged by an overweight truck but was restored and now open to traffic. It was named after the Kidwell Family who operated the Kidwell Grist and Saw Mills on the bank of Sunday Creek.

World Index Number: OH/35-05-02
National Register of Historic Places: April 11, 1977

Blackwood Covered Bridge

Township: Lodi
County: Athens

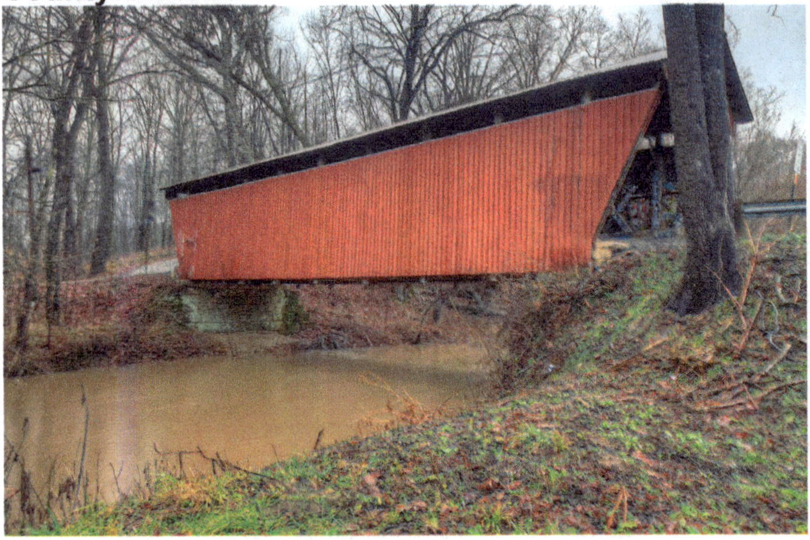

GPS Position: N 39° 11.848' W 81° 58.398'
Directions: From Garden head south on Garden Hill Rd for 0.5 mi and turn left onto Zion Rd. After 0.9 mi continue onto Blackwood Rd to find the bridge
Crosses: Shade River, Middle Branch
Carries: Blackwood Road
Builder: E.B.Henderson
Year Built: 1879 (R2003)
Truss Type: Multiple Kingpost
Dimensions: 1 Plus Span, 64 feet

Notes: Open to traffic, the 2003 renovation has left the bridge structurally in good shape but the graffiti has ruined its appearance. It is open for local traffic.

World Index Number: OH/35-05-06
National Register of Historic Places: June 23, 1978

Memorial Park (Saint Marys) Covered Bridge
Township: St. Marys
County: Auglaize

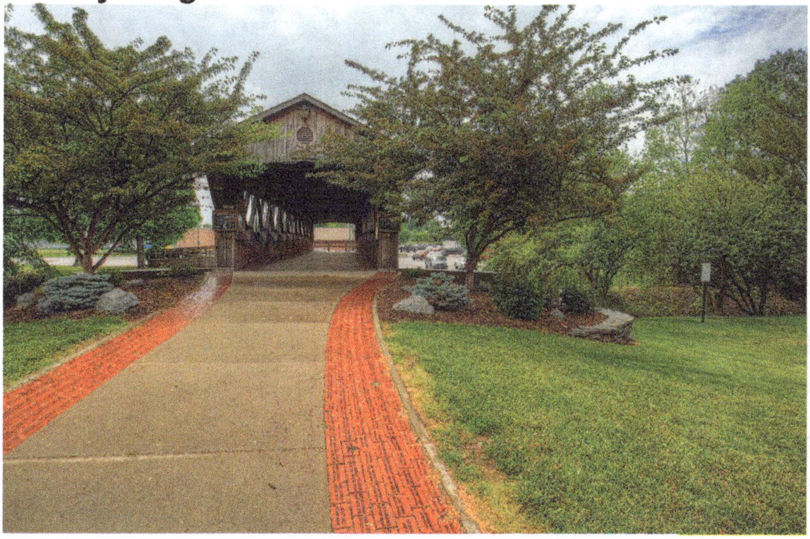

GPS Position: N 40° 32.614' W 84° 23.248'
Directions: Found in the town of St. Marys in a park accessed off East Spring St. west of Chestnut St.
Crosses: St. Marys River
Carries: Pedestrian walkway
Builder: Volunteers
Year Built: 1992
Truss Type: Howe
Dimensions: 1 Span, 105 feet

Notes: The shingles for the building were given by farmers in North Carolina in gratitude for the help they had received after a drought and hurricane. The wood for the shingles came from trees felled by the Hurricane.

World Index Number: OH/35-06-54
National Register of Historic Places: Not listed

Shaeffer (Campbell) Covered Bridge
Township: Richland
County: Belmont

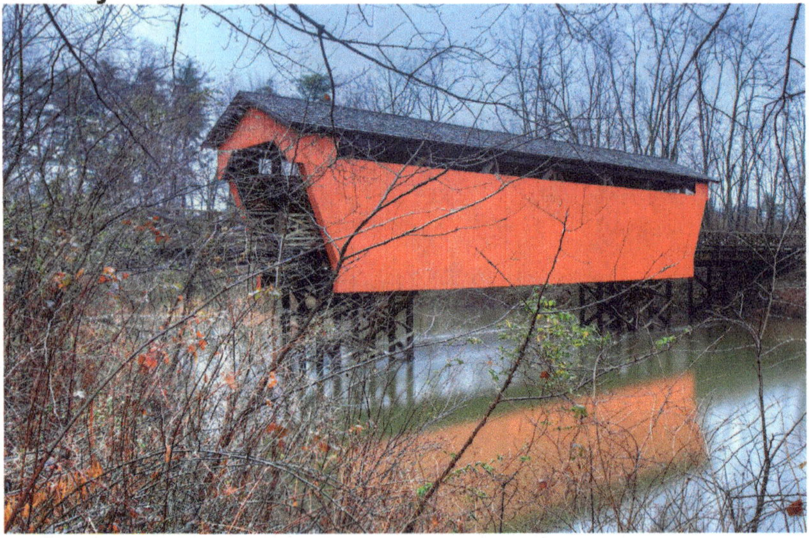

GPS Position: N 40° 04.479' W 80° 58.234'
Directions: Take OH-331 north from I-70 and after 0.2 miles turn left (East) on Township Rd 1518. The bridge is a short distance.
Crosses: Pond
Carries: Pedestrian Way
Builder: James W. Buchanan
Year Built: 1875 (M1973)
Truss Type: Multiple Kingpost
Dimensions: 1 Span, 68 feet
Notes: This bridge originally spanned Clear Creek in Fairfield County. After it collapsed under the weight of a farm tractor in 1973, it was gifted to the County of Belmont and was moved to its present location. It is in beautiful condition.

World Index Number: OH/35-07-05 originally OH/35-23-46
National Register of Historic Places: Not listed

Brown Covered Bridge

Township: Pike-Scott
County: Brown

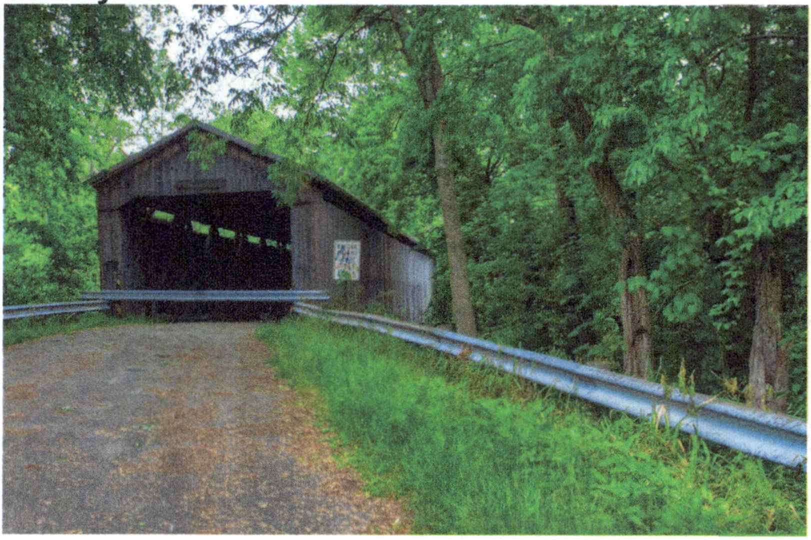

GPS Position: N 38° 59.438' W 83° 53.358'
Directions: From the town of Mt. Orab go east on Main St, continuing on CR-24B and after 1.9 miles turn right on New Hope-White Oak Station Rd and you will see the bridge in 1.9 miles.
Crosses: White Oak Creek
Carries: New Hope-White Oak Station Rd (Bypassed)
Builder: Smith Bridge Co.
Year Built: 1878 (R1960) (R2004)
Truss Type: Smith
Dimensions: 1 Span,129 feet

Notes: After damage from overloaded vehicles, it was closed and bypassed in 1999.

World Index Number: OH/35-08-04
National Register of Historic Places: Not listed

New Hope (Bethel Road) Covered Bridge

Township: Scott
County: Brown

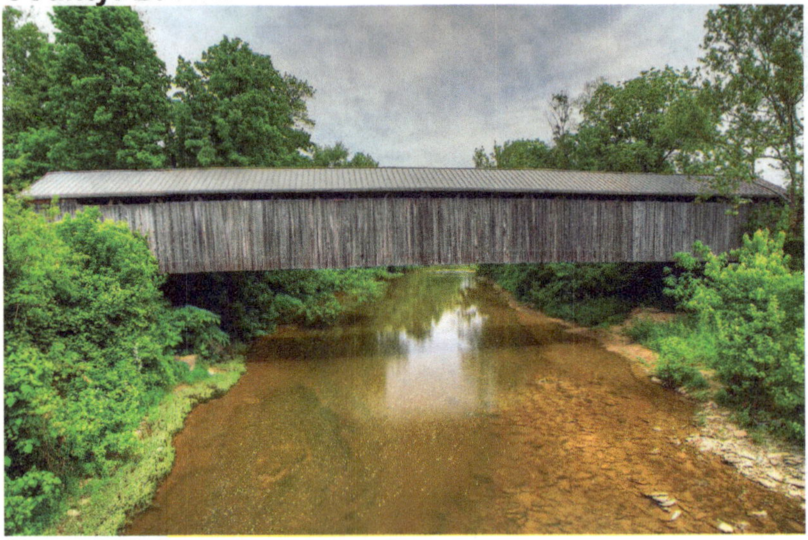

GPS Position: N 38° 57.576' W 83° 54.909'
Directions: From the town of Mt. Orab, go south on US-68 for 4.3 miles and turn right on Bethel-New Hope Rd where you will find the bridge on a bypassed section.
Crosses: White Oak Creek
Carries: Bethel-New Hope Rd (Bypassed Section)

Builder: Josiah Bryant
Year Built: 1878 (1902) (R1932) (R1977) (R2004)
Truss Type: Howe and Burr Arch
Dimensions: 1 Span, 170 feet

Notes: The laminated Burr Arches were added in 1902. It is the longest single span covered bridge in Ohio.

World Index Number: OH/35-08-05
National Register of Historic Places: Not listed

McCafferty Road Covered Bridge

Township: Perry
County: Brown

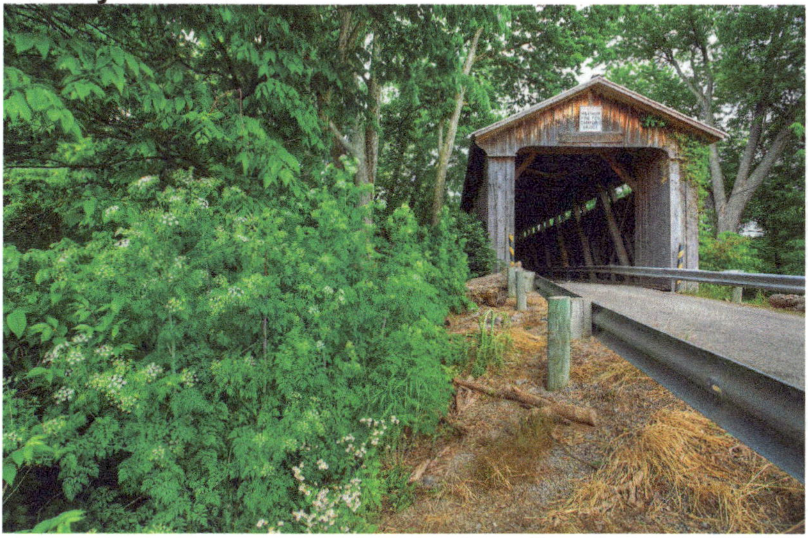

GPS Position: N 39° 09.100' W 83° 59.707'
Directions: From the town of Marathon, go northeast on US-50 for 0.6 miles and turn right on McCafferty Rd where you will see the bridge.
Crosses: Little Miami River, East Fork
Carries: McCafferty Rd
Builder: Smith Bridge Company
Year Built: 1877 (R1963)
Truss Type: Howe and Burr Arch
Dimensions: 1 Span, 157 feet

Notes: Open for traffic although it has been closed occasionally due to vehicle mishaps. It is unpainted and without windows.

World Index Number: OH/35-08-08
National Register of Historic Places: Not listed

North Pole Road Covered Bridge

Township: Huntington-Union
County: Brown

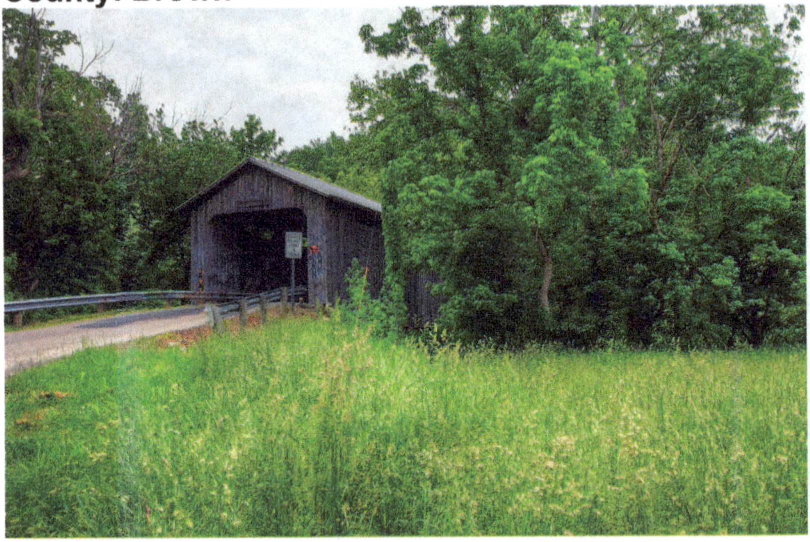

GPS Position: N 38° 45.029' W 83° 46.398'
Directions: From Ripley go east on US-62 for 0.8 miles and turn right on North Pole Rd where you will find the bridge in 3.0 miles.
Crosses: Eagle Creek
Carries: North Pole Rd (Bypassed in 2019)
Builder: Smith Bridge Co.
Year Built: 1875 (R1965) (R1997)
Truss Type: Smith
Dimensions: 1 Span, 156 feet
Notes: The 1997 repairs were required after the bridge was damaged by flood. An identical bridge upstream, the Eagle Creek Covered Bridge which was also built in 1875 by the same company, was destroyed in this flood.

World Index Number: OH/35-08-23
National Register of Historic Places: Not listed

George Miller Road Covered Bridge
Township: Byrd
County: Brown

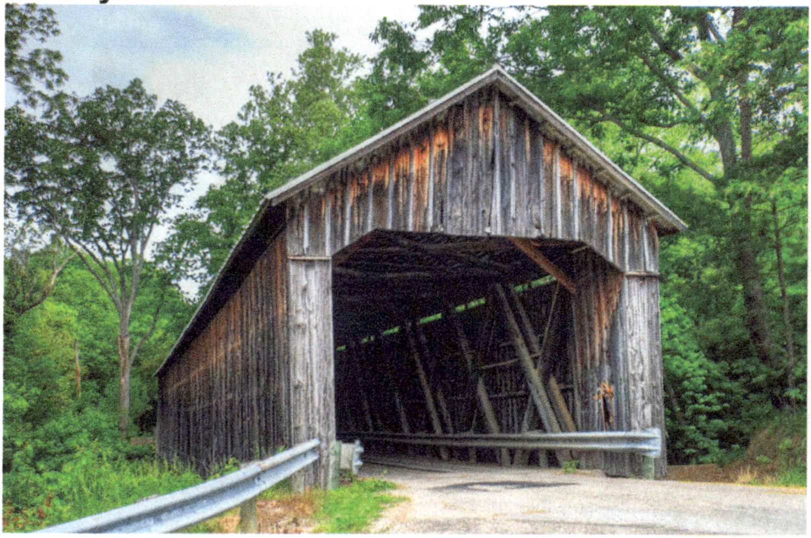

GPS Position: N 38° 50.203' W 83° 45.031'
Directions: From the town of Russellville go east on OH-125 for 3.9 miles and turn right on George Miller Rd where you will find the bridge in 1.2 miles.
Crosses: Eagle Creek, West Fork
Carries: George Miller Rd
Builder: Smith Bridge Co.
Year Built: 1878 (R1990s)
Truss Type: Smith Triple
Dimensions: 1 Span, 154 feet

Notes: The bridge is still active, In the 1990s it was renovated by local Amish and Mennonite crews from the Amos Schwartz Construction Company.

World Index Number: OH/35-08-34
National Register of Historic Places: Not listed

State Line (Bebb Park) Covered Bridge
Township: Morgan
County: Butler

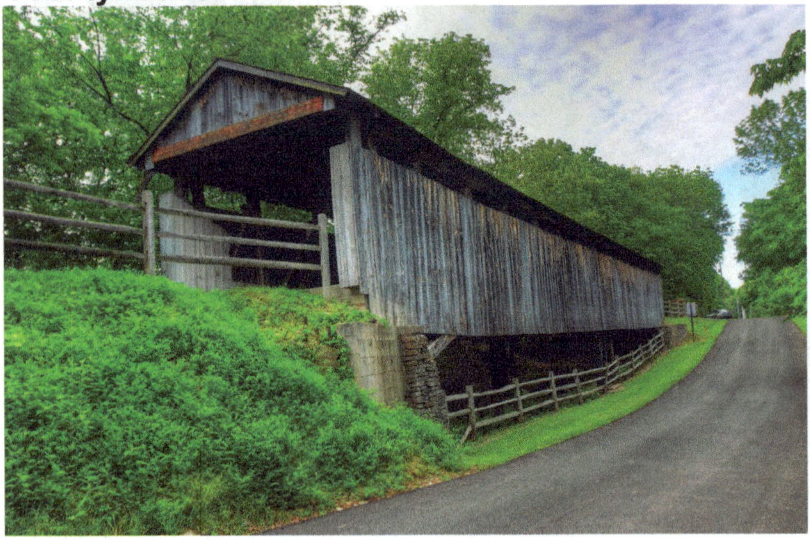

GPS Position: N 39° 22.273' W 84° 48.267'
Directions: From Milville, go west on OH-129 for 8.2 mi and turn left on OH-126/Cincinnati Brookville Rd. and go 1.0 mi Turn right on Bebb Park Ln and the bridge in 0.7 miles.
Crosses: Ravine
Carries: Bebb Park Ln
Builder: Banden, Butin & Bowman
Year Built: Ca. 1867 (M1886) (M1966)
Truss Type: Wernwag
Dimensions: 1 Span, 120 feet
Notes: This bridge has the only surviving example of the Wernwag truss, designed by the German engineer, Louis Wernwag.

World Index Number: OH/35-09-02
National Register of Historic Places: Not listed

Pugh's Mill (Black) Covered Bridge
Township: Oxford
County: Butler

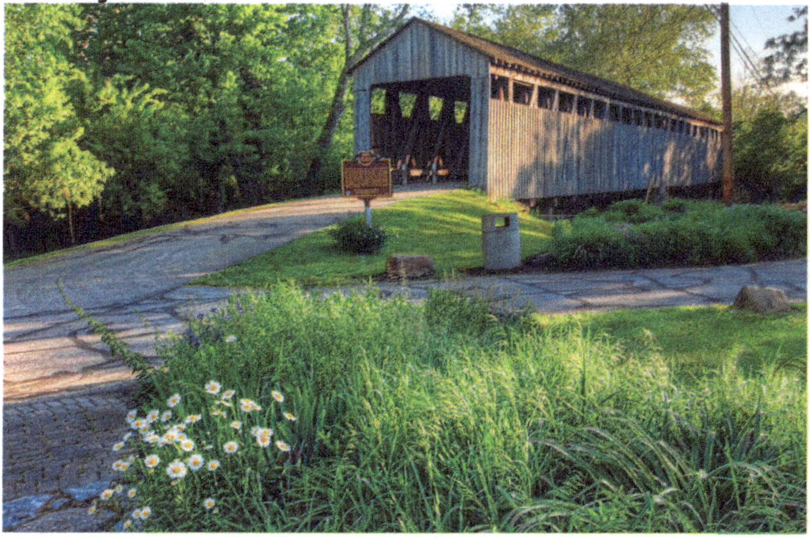

GPS Position: N 39° 31.405' W 84° 44.107'
Directions: In Oxford, go NW on OH-732/Morning Sun Rd. and after 0.1 mi, turn left on Corso Rd and the bridge is a short distance
Crosses: Four Mile Creek
Carries: Corso Rd (Bypassed Section)
Builder: Bandin, Butin and Bowman
Year Built: 1868 (R1869) (R2000)
Truss Type: Long
Dimensions: 1plus Span, 209 feet
Notes: Built to give access to a saw and grist mill owned by James B. Pugh on Four Mile (Tallawanda) Creek. It was modified to a two span from one, shortly after construction. It was bypassed by a new bridge in the 1950s.

World Index Number: OH/35-09-03
National Register of Historic Places: June 5, 1975

Blue Bird Farm (Indian Fork) Covered Bridge
Township: Center
County: Carroll

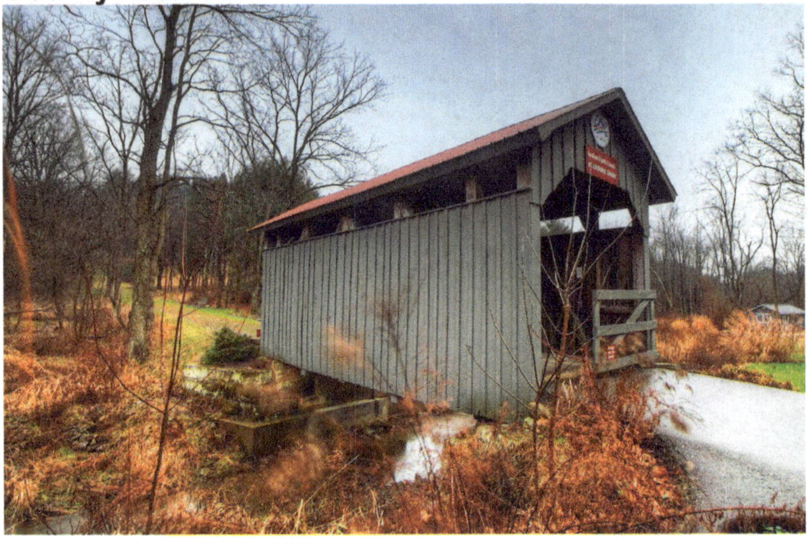

GPS Position: N 40° 34.131' W 81° 04.801'
Directions: In the southeast portion of the town of Carrollton near 198 Alamo Rd.
Crosses: Indian Fork River
Carries: Park Trail

Builder: Jim Farley
Year Built: 1996
Truss Type: Multiple Kingpost
Dimensions: 1 Span, 30 feet

Notes: On private property but accessible. Carries a pedestrian trail over Indian Creek.

World Index Number: OH/35-10-14
National Register of Historic Places: Not listed

Perintown (Stonelick) Covered Bridge
Township: Stonelick
County: Cleremont

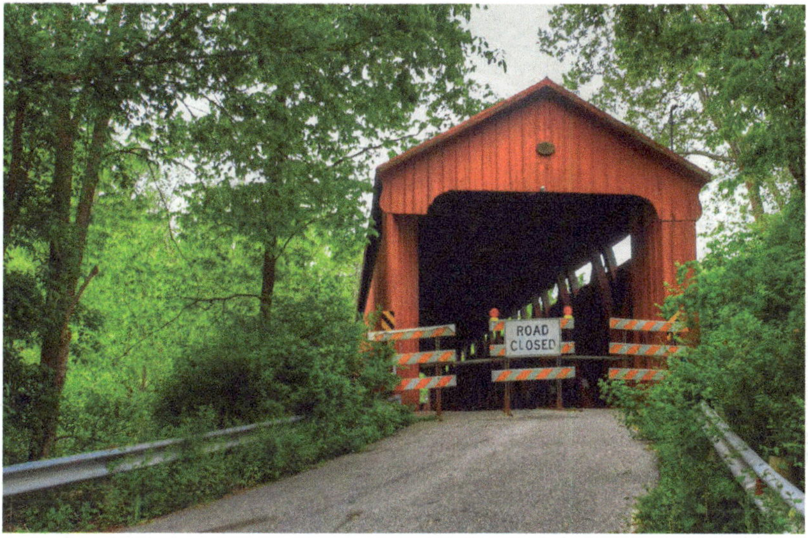

GPS Position: N 39° 07.862' W 84° 11.226'
Directions: From the town of Owensville, go west on US-50 for 3.2 miles and turn right on Stonelick Williams Corner Rd where you will find the bridge in 1.2 miles.
Crosses: Stonelick Creek
Carries: Stonelick Williams Corner Rd
Builder: M. Collins
Year Built: 1878 (R1971) (R2007) (R2015)
Truss Type: Howe
Dimensions: 1 Span, 140 feet

Notes: The bridge was damaged in June 2010 by an overloaded truck but re-opened in 2015

World Index Number: OH/35-13-02
National Register of Historic Places: September 10, 1974

Martinsville (Cuba Pike) Covered Bridge

Township: Clark
County: Clinton

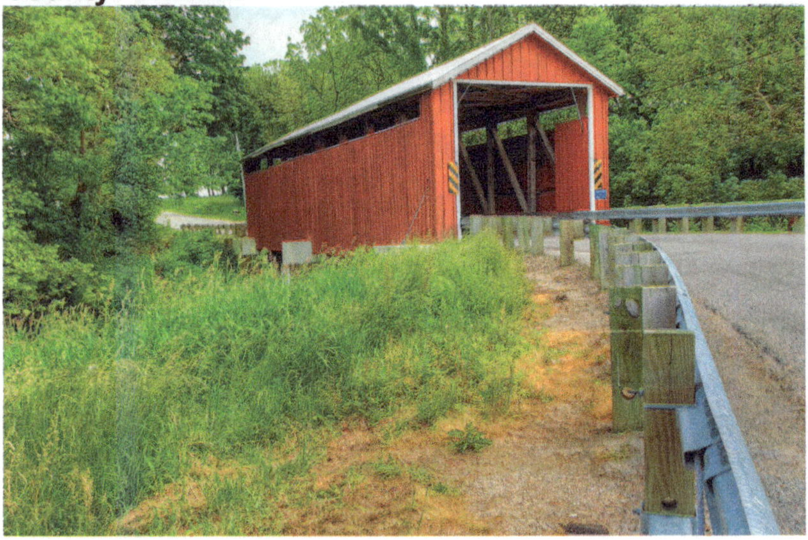

GPS Position: N 39° 19.816' W 83° 50.112'
Directions: From Martinsville, go west on OH-28 for 0.5 mi and turn right on Martinsville Rd where you will find the bridge in 0.7 miles.
Crosses: Todd's Fork
Carries: Martinsville Rd
Builder: Zimri Wall and Champion Bridge Company
Year Built: 1871 (R1970s) (R1980s) (R2006)
Truss Type: Multiple Kingpost
Dimensions: 1 plus Span, 80 feet
Notes: This bridge still carries traffic. It has been reinforced with concrete abutments as well as steel I-beams. In the 2006 renovation the roof was raised to provide a height clearance of 14' 6" from 11' 6".

World Index Number: OH/35-14-09
National Register of Historic Places: September 10, 1974

Lynchburg Covered Bridge
Township: Clark-Dodson
County: Clinton-Highland

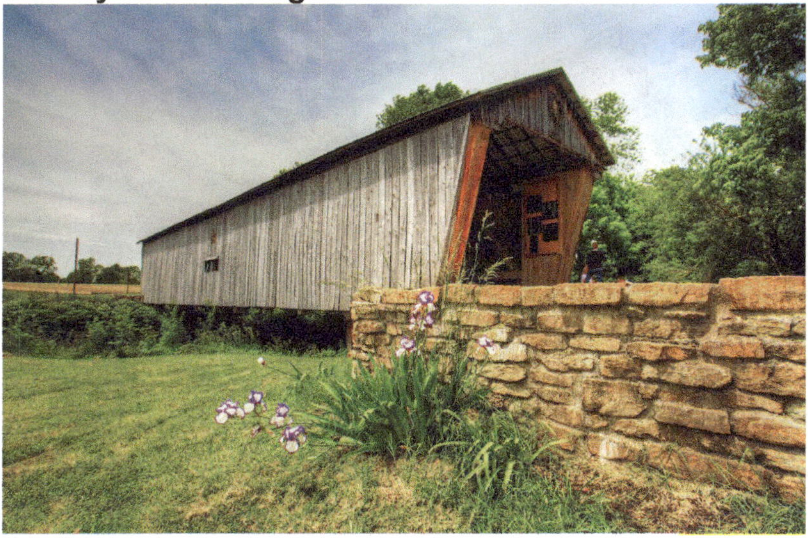

GPS Position: N 39° 14.708' W 83° 47.637'
Directions: On a bypassed section of road at the west end of High St. in the town of Lynchburg.
Crosses: Little Miami River, East Fork
Carries: N/A
Builder: John C. Gregg
Year Built: 1870
Truss Type: Long
Dimensions: 1 Span, 120 feet

Notes: The bridge has an unusual metal truss which looks to be an inverted bowstring and is probably a later addition to the Long Truss. The bridge was slated for demolition in 1963 but community support helped save it.

World Index Number: OH/35-14-11 and OH/35-36-06
National Register of Historic Places: 03/16/1976

Sells (Roller Mill) Covered Bridge

Township: Center
County: Columbiana

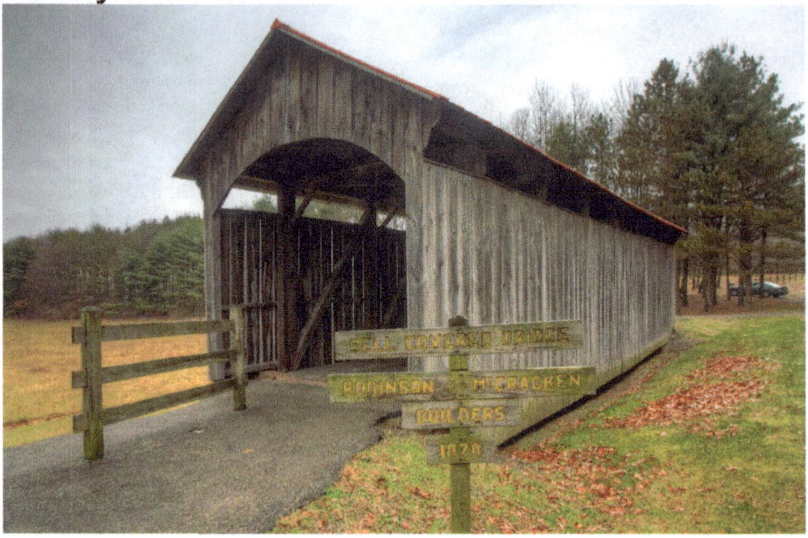

GPS Position: N 40° 44.319' W 80° 49.197'
Directions: From Lisbon head west on US-30 W/W Lincoln Way for 1.1 mi and turn left onto Lisbon-Millport Rd. After 3.0 mi turn right into Scenic Vista Park and the bridge is 0.2 mi
Crosses: None
Carries: Park Road
Builder: Robinson and McCracken (Original)
Year Built: 1995 (Original 1878)
Truss Type: Multiple Kingpost
Dimensions: 1 Span, 48 feet
Notes: The original bridge spanned Little Beaver Creek. It was removed in 1992 and dismantled. It was rebuilt in its present location but was destroyed by arson on July 19, 1993. The present structure was completed in 1995.

World Index Number: OH/35-15-01#2
National Register of Historic Places: Not listed

Jim McClellan Covered Bridge

Township: Center
County: Columbiana

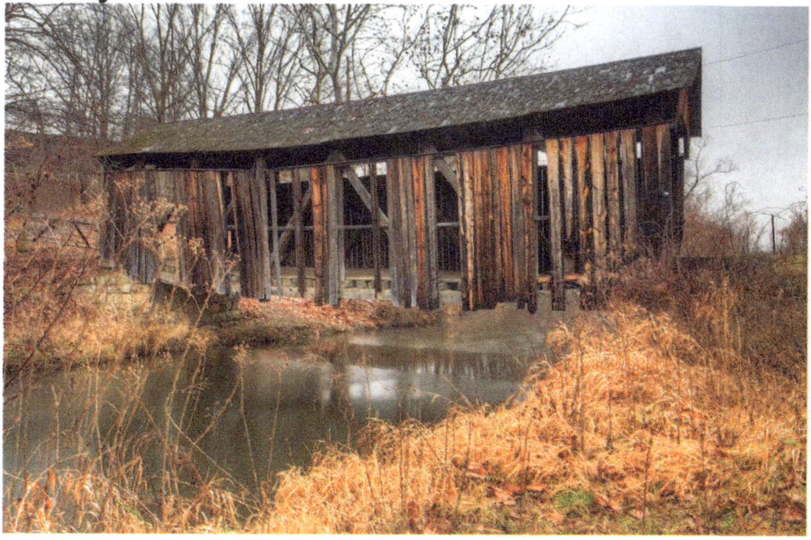

GPS Position: N 40° 43.734' W 80° 50.186'
Directions: From Hanoverton go east on U.S. 30 for 3.8 miles and turn right on trinity Church Rd. After 3.4 miles you will see an old road to the right where the bridge is a short distance.
Crosses: Little Beaver Creek, West Fork
Carries: Abandoned road

Builder: Not known
Year Built: 2016
Truss Type: Multiple Kingpost
Dimensions: 1 Span, 53 feet
Notes: The original bridge was in terrible condition (see image above) and was replaced in 2016. There was a cane mill just north of the bridge.

World Index Number: OH/35-15-02#2
National Register of Historic Places: Not listed

Teagarden (Centennial) Covered Bridge
Township: Salem
County: Columbiana

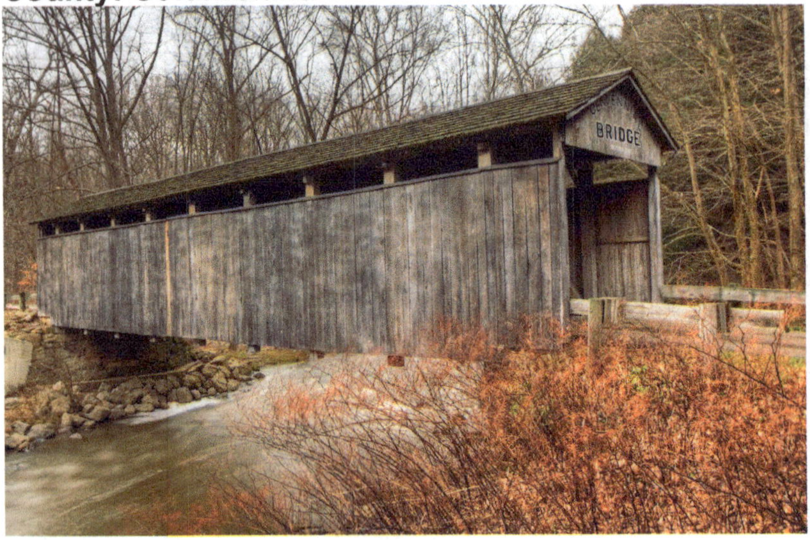

GPS Position: N 40° 49.318' W 80° 49.631'
Directions: From Lisbon, go northwest on OH-45 and after 2.9 miles go left on St Jacobs Logtown Rd. Proceed for 0.7 miles and turn right on Eagleton Rd where you will find the bridge in 1.2 miles.
Crosses: Little Beaver Creek, Middle Fork
Carries: Eagleton Rd.

Builder: J. C. Mountz
Year Built: 2003 (Original 1875)
Truss Type: Multiple Kingpost
Dimensions: 1 Span, 65 feet
Notes: The original bridge suffered fire damage and was replaced in 2003.

World Index Number: OH/35-15-05#2
National Register of Historic Places: Not listed

Church Hill Road Covered Bridge
Township: Elk Run
County: Columbiana

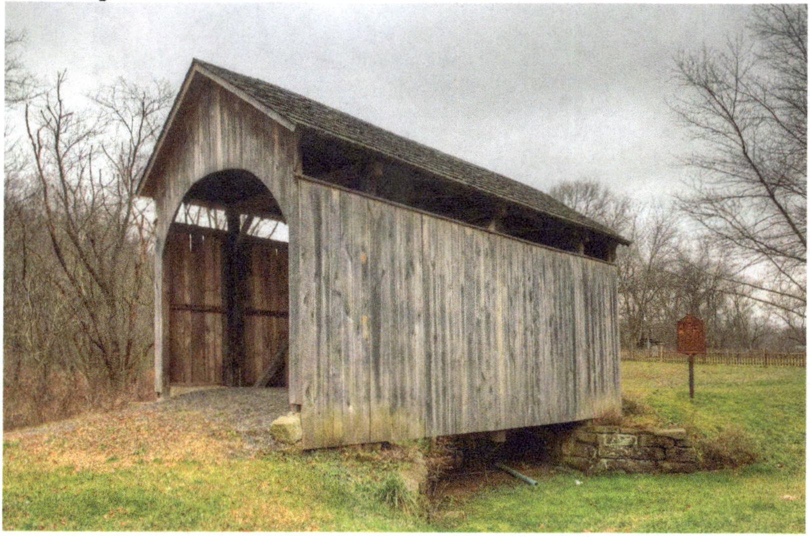

GPS Position: N 40° 45.726' W 80° 42.231'
Directions: From Lisbon go east on OH-154/ Middle Beaver Rd for 1.4 miles where you will see the bridge by the side of the road.
Crosses: None
Carries: OH-154
Builder: Not known
Year Built: 1870s (M1982)
Truss Type: King
Dimensions: 1 Span, 19 feet

Notes: It was originally located over Middle Fork Little Beaver Creek in Elk Township but was moved to its present location in 1982. It suffered from a fire in 2003.

World Index Number: OH/35-15-08
National Register of Historic Places: June 11/ 1975

Thomas J. Malone Covered Bridge

Township: St. Clair
County: Columbiana

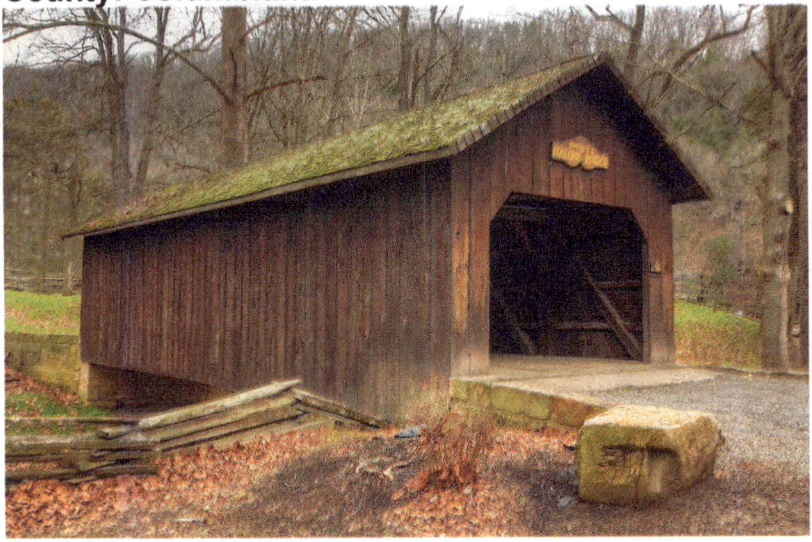

GPS Position: N 40° 43.616' W 80° 36.735'
Directions: From the village of Clarkson, go south on Echo Dell Rd for 1.5 miles where you will see the bridge.
Crosses: Gaston Mill Tailrace
Carries: Echo Dell Rd
Builder: Not known
Year Built: 1870 (M CA.1900) (M CA.1912) (M CA.1971) (R1971)
Truss Type: Multiple Kingpost
Dimensions: 1 Span, 42 feet
Notes: About 1912 it was moved to Pine Hollow Road where it was used for storage. The bridge is named after a researcher who recognized this structure as a covered bridge.

World Index Number: OH/35-15-96
National Register of Historic Places: Not listed

Helmick Covered Bridge

Township: Clark
County: Coshocton

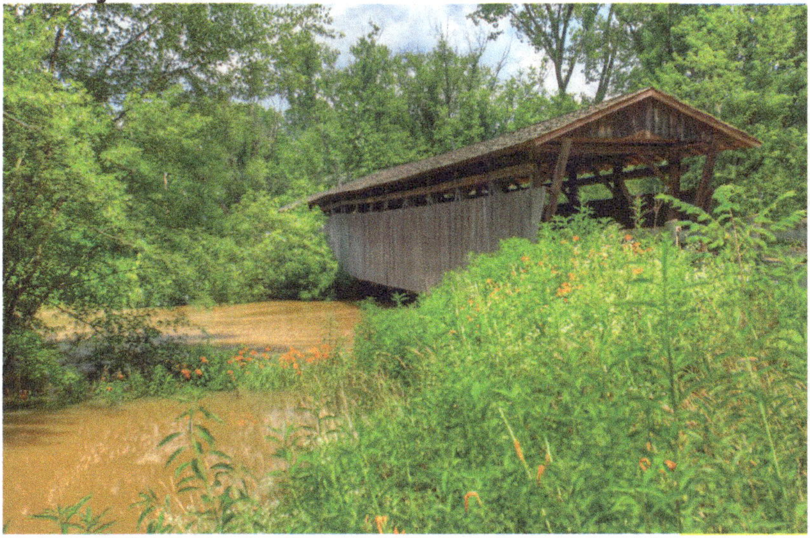

GPS Position: N 40° 23.604' W 81° 56.552'
Directions: From Blissfield go east on CR-343 for 1.4 miles and turn left on Township Rd 26 where the site is a short distance.
Crosses: Killbuck Creek
Carries: Township Rd 26
Builder: John Shrake built the original bridge.
Year Built: 1996 (Original 1863)
Truss Type: Multiple Kingpost
Dimensions: 2 Span, 166 feet
Notes: The first bridge was closed in 1980 and taken apart in 1995. About 20% of the timbers were used in the replacement bridge. Not to be confused with the Helmick Mill Covered Bridge, in Morgan County, Ohio.

World Index Number: OH/35-16-02#2
National Register of Historic Places: Not listed

Charles A. Harding Memorial Covered Bridge
Township: Olmsted
County: Cuyahoga

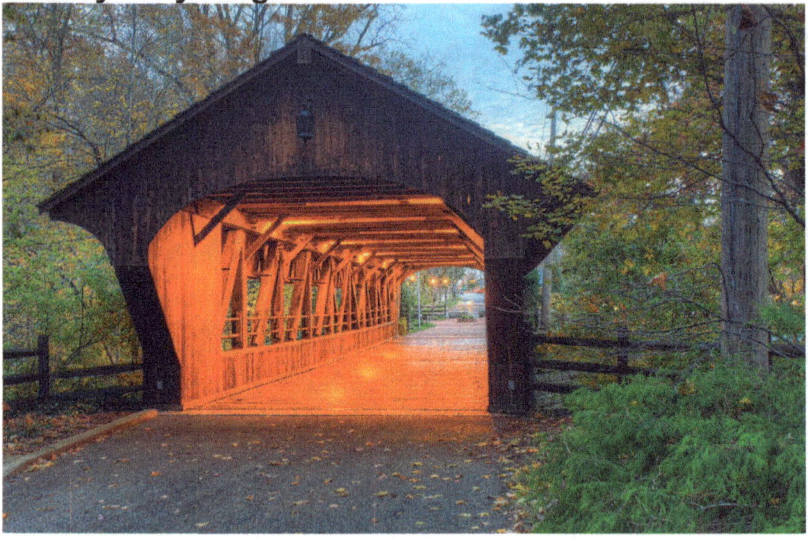

GPS Position: N 41° 22.668' W 81° 54.071'
Directions: In the town of Olmsted Falls, found off the south end of Main St. south of Water St. Pedestrians only
Crosses: Plum Creek
Carries: Pedestrian walkway
Builder: Richland Engineering
Year Built: 1998
Truss Type: Multiple Kingpost
Dimensions: 1 Span, 92 feet

Notes: The bridge was named after Captain Charles Harding (1915-1944), from Olmsted Falls, who was killed in action during World War II in the battle of Normandy. It was built for pedestrian use only.

World Index Number: OH/35-18-25
National Register of Historic Places: Not listed

Chambers Road Covered Bridge

Township: Porter
County: Delaware

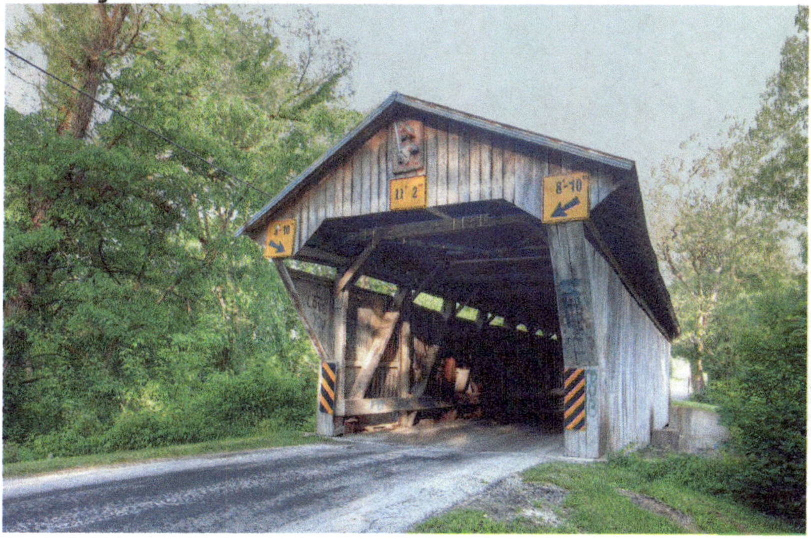

GPS Position: N 40° 20.298' W 82° 49.106'
Directions: From Olive Green, go northeast on OH-656 for 0.6 miles and turn left on Hicks Rd./Liberty Rd. N. After 0.8 miles turn left on Chambers Rd. where you will find the bridge in 0.2 miles.
Crosses: Big Walnut Creek
Carries: Chambers Road
Builder: Everett S. Sherman
Year Built: 1883 (R1957) (R1982)
Truss Type: Childs
Dimensions: 1 plus Span, 74 feet

Notes: This one lane bridge is still in use and is the last surviving covered bridge in Delaware County.

World Index Number: OH/35-21-04
National Register of Historic Places: November 21, 1974

Hizey (Visitine) Covered Bridge
Township: Violet
County: Fairfield

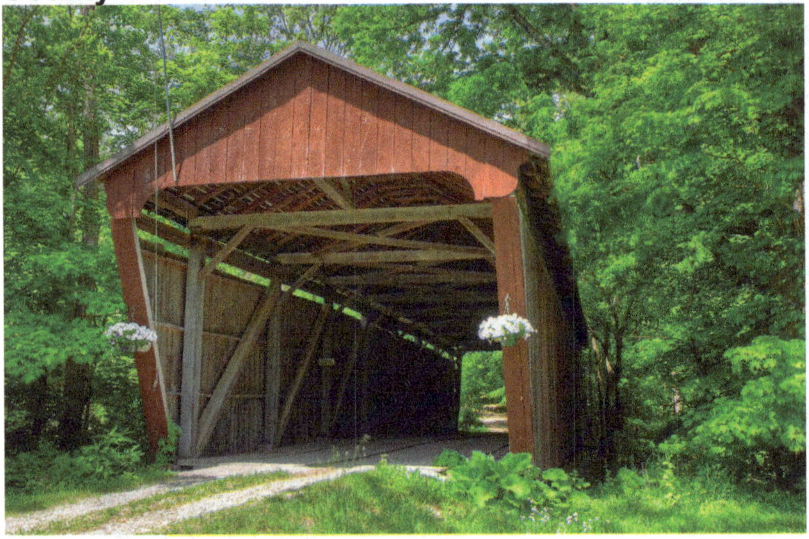

GPS Position: N 39° 54.866' W 82° 42.528'
Directions: In the town of Pickerton found by Toll Gate Road 0.6 miles north of Refuge Road on private property.
Crosses: Sycamore creek
Carries: Private Road
Builder: J.W. Buchanan
Year Built: 1891 (R1914) (M1986) (R1989)
Truss Type: Multiple Kingpost and Queen
Dimensions: 1 Span, 83 feet

Notes: The original site was crossing Poplar Creek on Poplar Creek Road, it was moved to its present location in 1979. Burr arches were added in 1914 but when the bridge was rebuilt at the present site in 1986, they were left off.

World Index Number: OH/35-23-07
National Register of Historic Places: October 8, 1976

John Bright #2 Covered Bridge
Township: Lancaster
County: Fairfield

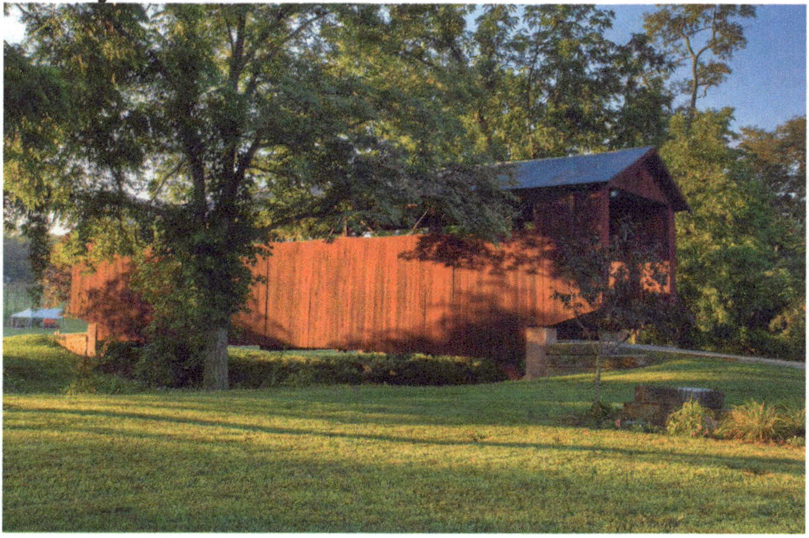

GPS Position: N 39° 44.361' W 82° 35.123'
Directions: In the north side of the town of Lancaster, found on the campus of Ohio University by Granville Pike/OH-37
Crosses: Fetters Run
Carries: None
Builder: August Borneman
Year Built: 1881 (M1988) (R1988)
Truss Type: Inverted Bowstring with Arch
Dimensions: 1 Span, 75 feet

Notes: This is one of only two bridges which use the metal Inverted Bowstring Truss. The arch is a later addition. The bridge originally crossed the Poplar Creek on Township Road 263 and was moved and renovated in its current site in 1988.

World Index Number: OH/35-23-10
National Register of Historic Places: May 28, 1975

George Hutchins Covered Bridge
Township: Lancaster
County: Fairfield

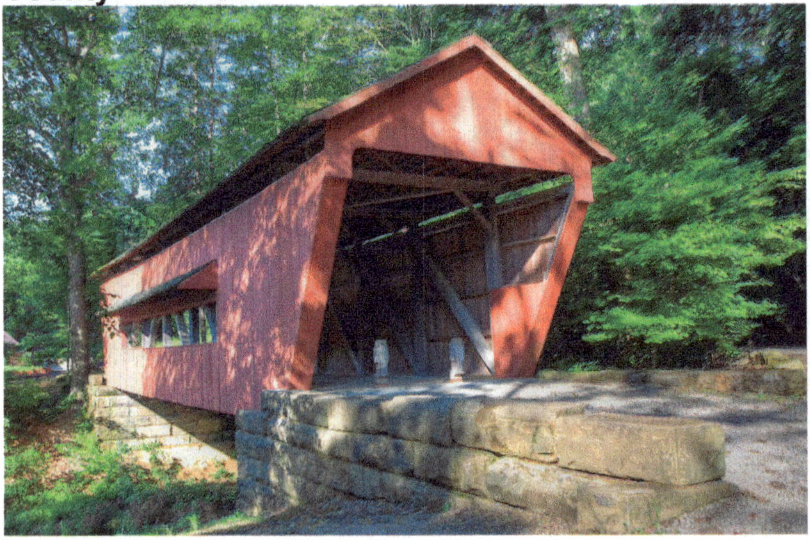

GPS Position: N 39° 40.849' W 82° 34.712'
Directions: Found in Alley Park off Old Logan Road, 1.2 miles south of Lancaster
Crosses: Lake Loretta
Carries: Pedestrian walkway

Builder: J.W.Buchanan
Year Built: 1904 (M1988) (R2000)
Truss Type: Multiple Kingpost
Dimensions: 1 Span, 62 feet

Notes: Originally built 1904 over Clear Creek on Strickler Road near Amanda, Ohio. It was dismantled in 1988,and relocated and renovated in 2000.

World Index Number: OH/35-23-13
National Register of Historic Places: Not listed

Hanaway (Clearport) Covered Bridge

Township: Madison
County: Fairfield

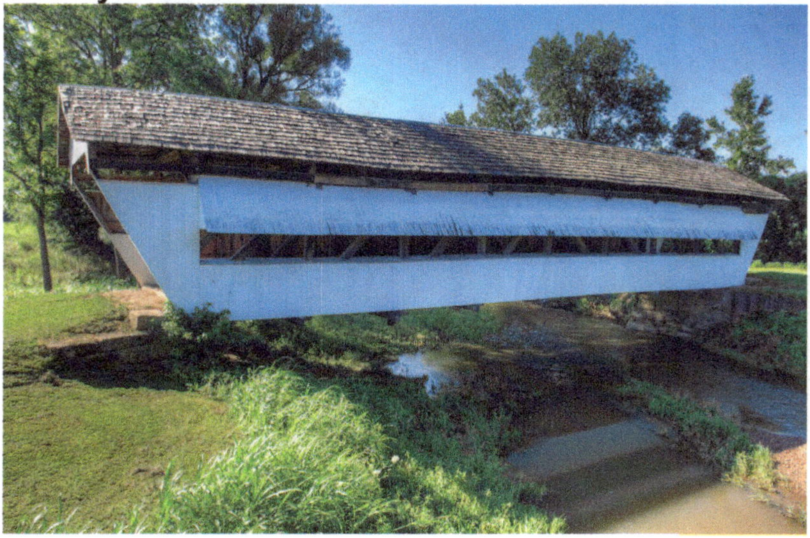

GPS Position: N 39° 37.091' W 82° 40.868'
Directions: From Amanda, go east on Hamburg Road for 0.3 miles and turn left on OH-159/Chiiicothe-Lancaster Rd. and then in 0.1 miles, turn right on Hamburg Rd In 2.7 miles turn right on Clearport Road and find the bridge on a bypassed section in 2.1 miles.
Crosses: Clear Creek
Carries: Clearport Road (Bypassed Section)
Builder: J.W.Buchanan
Year Built: 1901 (R2002)
Truss Type: Multiple Kingpost
Dimensions: 1 Span, 86 feet
Notes: The bridge was bypassed in 1990 and had deteriorated to a great extent before the 2002 renovation.

World Index Number: OH/23-15
National Register of Historic Places: Not listed

Johnston (Terry Mill) Covered Bridge

Township: Madison
County: Fairfield

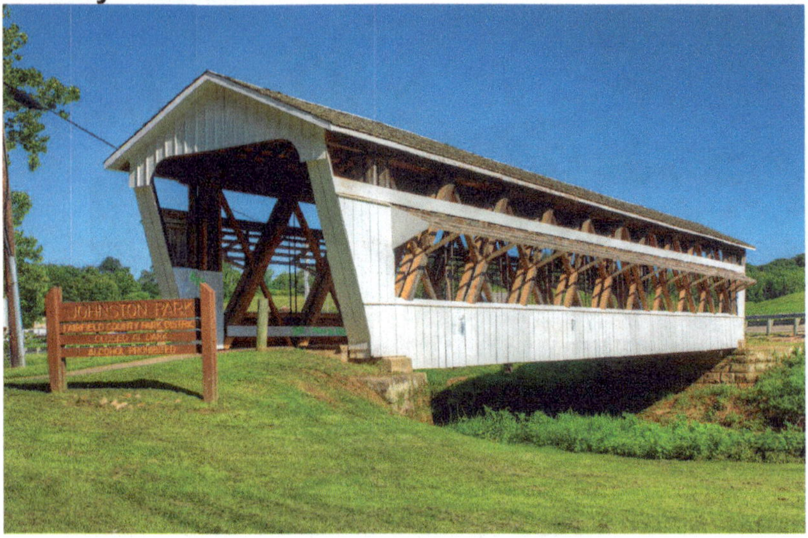

GPS Position: N 39° 36.778' W 82° 39.511'
Directions: From Lancaster, go SW on Boving Rd for 1.9 mi and turn right on Hamburg Rd. Go 2.7 mi and turn left on Hopewell Church Rd. After 2.6 miles turn left on Clearcreek Road where you will find the bridge in 0.6 miles
Crosses: Clear Creek
Carries: Clearcreek Road (Bypassed section)
Builder: August Borneman
Year Built: 1887
Truss Type: Howe
Dimensions: 1 Span, 98 feet
Notes: Closed and bypassed in 1990, it looks in wonderful shape and must have had a recent renovation. The high bridge length windows look great, inside and out

World Index Number: OH/35-23-16
National Register of Historic Places: **Not listed**

Zeller-Smith (Sycamore) Covered Bridge
Township: Violet
County: Fairfield

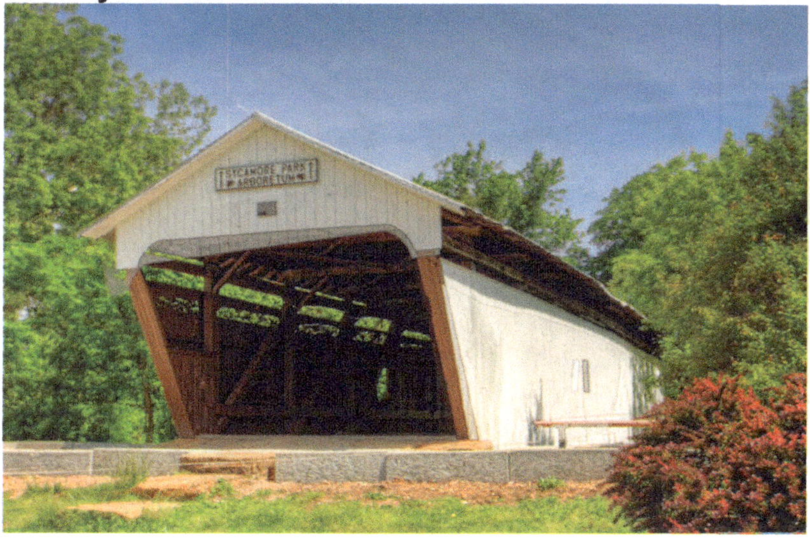

GPS Position: N 39° 52.814' W 82° 44.882'
Directions: In the town of Pickerington, go east on Covered Bridge Ln off Pickerington Road and you will find the bridge in a short distance in Sycamore Park.
Crosses: Sycamore Creek
Carries: None
Builder: J.W.Buchanan
Year Built: Ca.1905 (M1986)
Truss Type: Multiple Kingpost
Dimensions: 1 Span, 79 feet
Notes: The bridge originally crossed Sycamore Creek on Busey Road two miles south of Pickerington, Ohio. After it was closed in 1984, it was moved to the park in 1986. It is in a nice setting with wildflowers growing along the Creek.

World Index Number: OH/35-23-19
National Register of Historic Places: Not listed

McCleery Covered Bridge
Township: Lancaster
County: Fairfield

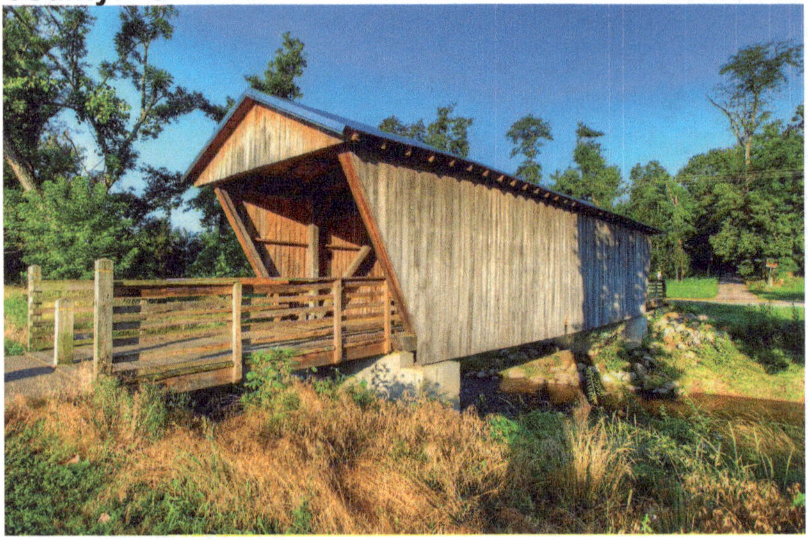

GPS Position: N 39° 43.498' W 82° 35.127'
Directions: Located in a park at the north end of Lancaster by Arbor Valley Trail north of Fair Ave.
Crosses: Fetter's Run
Carries: Bicycle path
Builder: Jacob R. Brandt
Year Built: 1864 (M1983) (M2006)
Truss Type: Multiple Kingpost
Dimensions: 1 Span, 52 feet
Notes: The bridge was initially built over Walnut Creek south of Baltimore, Ohio and in 1983 was moved to an open field on private property. In 2006 it was moved to its present site as part of a bike path.

World Index Number: OH/35-23-25
National Register of Historic Places: Not listed

Shryer (Game Farm) Covered Bridge

Township: Liberty
County: Fairfield

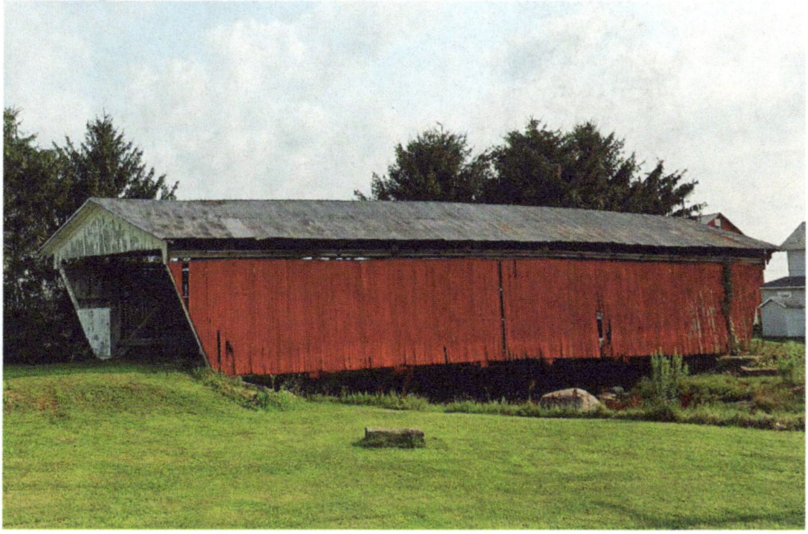

GPS Position: N 39° 50.731' W 82° 40.686'
Directions: From the town of Baltimore, go west on Basil Western Rd. for 2.9 miles where you will find the bridge beside the road.
Crosses: Ditch
Carries: Beside Basil Western Rd.
Builder: J.W.Buchanan
Year Built: 1891 (M1987)
Truss Type: Multiple Kingpost
Dimensions: 1 Span, 66 feet
Notes: The bridge was moved to its present location in 1987 and is now used as storage. It has an interesting skewed portal. It is called Shryer Covered Bridge but the family that owns the farm spells their name Schreyer.

World Index Number: OH/35-23-27
National Register of Historic Places: Not listed

Andrews (Charles Holliday) Covered Bridge
Township: Walnut
County: Fairfield

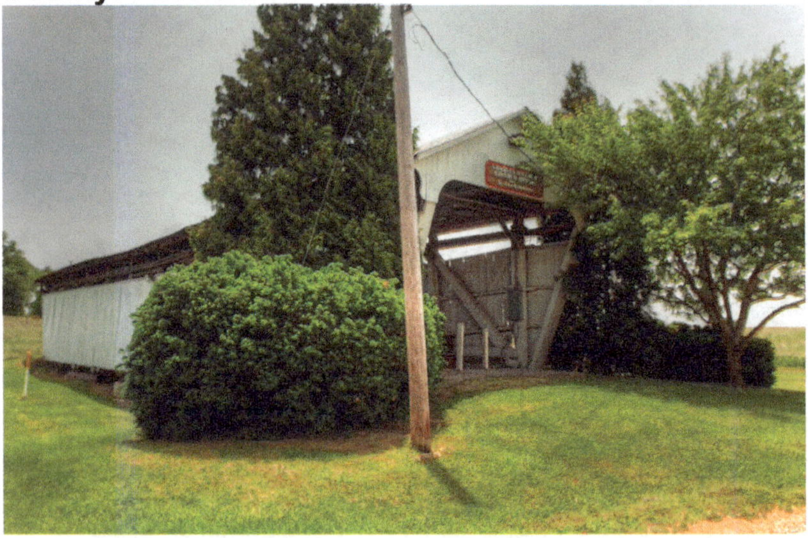

GPS Position: N 39° 53.789' W 82° 31.577'
Directions: Found in the town of Millersport on the Corn Festival grounds off Chatauqua Blvd.
Crosses: None
Carries: N/A
Builder: J.W.Buchanan
Year Built: 1897 (M1982)
Truss Type: Multiple Kingpost
Dimensions: 1 Span, 98 feet

Notes: It originally carried traffic over Walnut Creek but was closed in 1982 and given to the Corn Festival. It was originally built using some of the remnants of an 1891 bridge that once stood at the original site.

World Index Number: OH/35-23-30
National Register of Historic Places: Not listed

R.F. Baker (Winegardner) Covered Bridge
Township: Richland
County: Fairfield

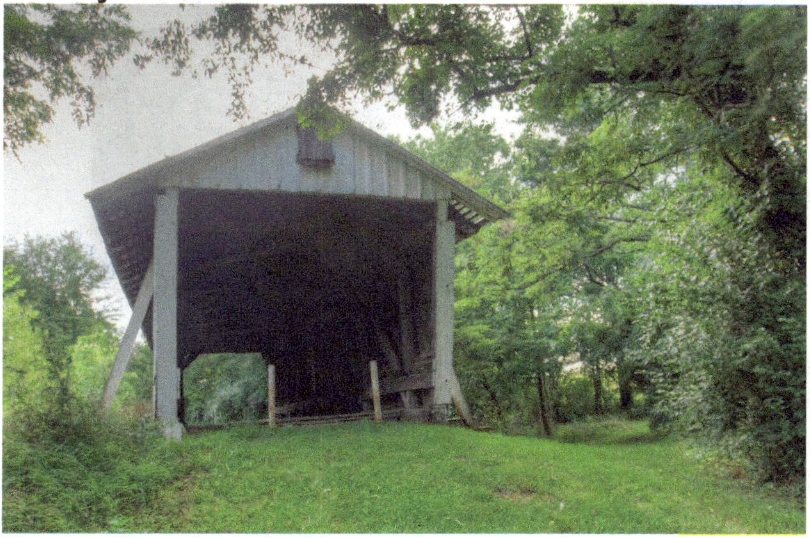

GPS Position: N 39° 45.117' W 82° 28.278'
Directions: Found behind the Sports Field of Fairfield Union High School, 1.3 miles west of Rushville on US-22
Crosses: Pond
Carries: Pedestrian path
Builder: James Arnold
Year Built: 1871 (M1981)
Truss Type: Multiple Kingpost
Dimensions: 1 Span, 66 feet
Notes: The bridge initially spanned Little Rush Creek but was moved in 1981 when a reservoir was built. It stood near to where the Winegardner Mill once stood, hence its alternate name. It seems a bit forgotten and run down in its present location.

World Index Number: OH/35-23-33
National Register of Historic Places: Not listed

Jon Raab Covered Bridge

Township: Pleasant
County: Fairfield

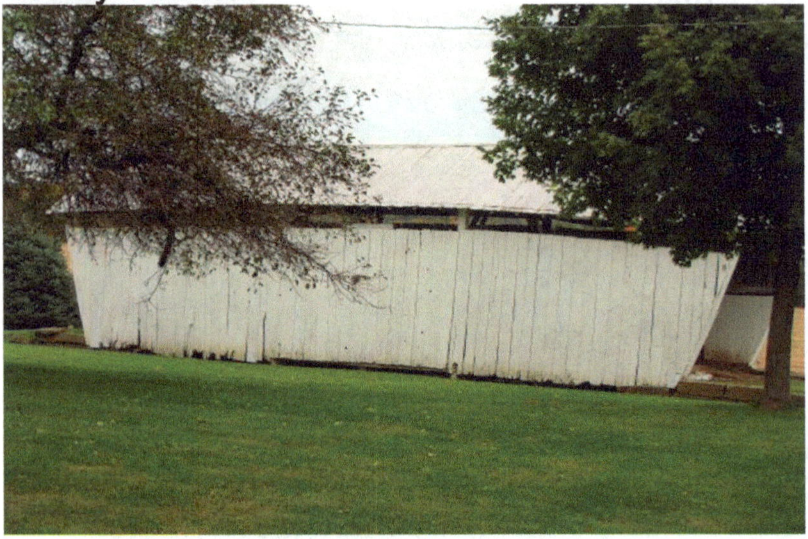

GPS Position: N 39° 43.829' W 82° 29.209'
Directions: From Lancaster go east on US-22 for 1.5 miles and turn on Ireland Rd north. The bridge is on private property in 1.2 miles
Crosses: None
Carries: None
Builder: William Funk
Year Built: 1891 (M1974)
Truss Type: Queen
Dimensions: 1 Span, 44 feet

Notes: The bridge originally spanned Raccoon Run on Ireland Road near Rushville but was moved to its present location in 1974 where it is being used as a storage shed.

World Index Number: OH/35-23-37
National Register of Historic Places: Not listed

Hartman #2 (Lockville Park) Covered Bridge

Township: Violet
County: Fairfield

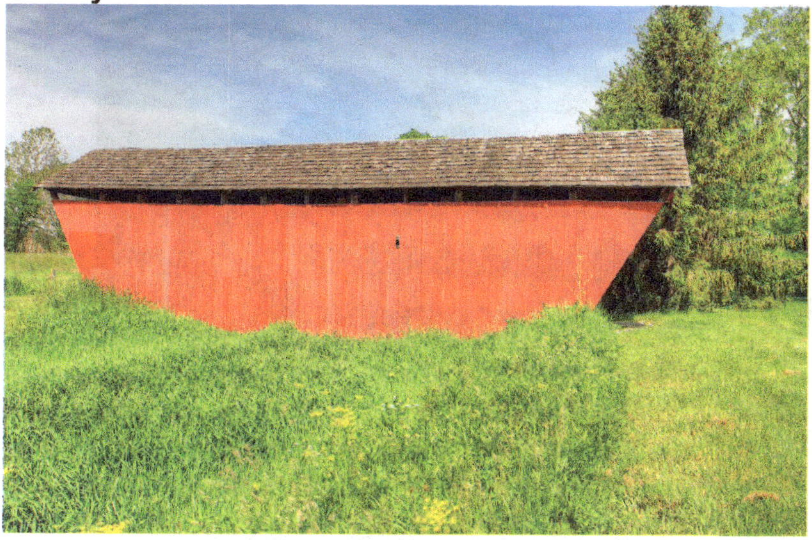

GPS Position: N 39° 49.066' W 82° 44.216'
Directions: From Carroll, go west on Winchester Road for 1.4 mi and turn right on Pickerington Rd where the bridge is 1.2 miles.
Crosses: None
Carries: None
Builder: William Funk
Year Built: 1888 (R1967) (M1967)
Truss Type: Queen
Dimensions: 1 Span, 48 feet

Notes: The bridge originally crossed Pleasant Run carrying Wheeler Road in Lancaster, Ohio. It was moved in 1967 across some remains of the Ohio-Erie Canal.

World Index Number: OH/35-23-38
National Register of Historic Places: Not listed

Mink Hollow Covered Bridge

Township: Hocking
County: Fairfield

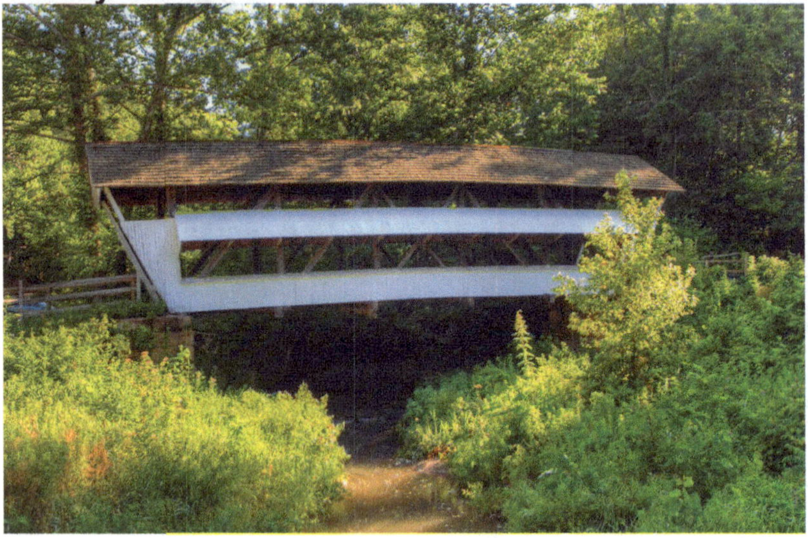

GPS Position: N 39° 38.350' W 82° 39.002'
Directions: From Ancaster, go south on Boving Rd for 1.4 mi and turn right on Hamburg Rd. After 2.7 miles turn left on Hopewell Church Rd and then after 2.2 miles turn left on Crooks Road. You will find the bridge in 2.1 miles.
Crosses: Arney Run
Carries: Crooks Road (Bypassed section)
Builder: J.R.Brandt
Year Built: 1887 (R1996) (R2012)
Truss Type: Multiple Kingpost
Dimensions: 1 Span, 51 feet

Notes: The Multiple Kingpost truss is a variation on the normal design, having a centred x-shape. It is illuminated at night

World Index Number: OH/35-23-43
National Register of Historic Places: Not listed

Rock Mill Covered Bridge
Township: Bloom
County: Fairfield

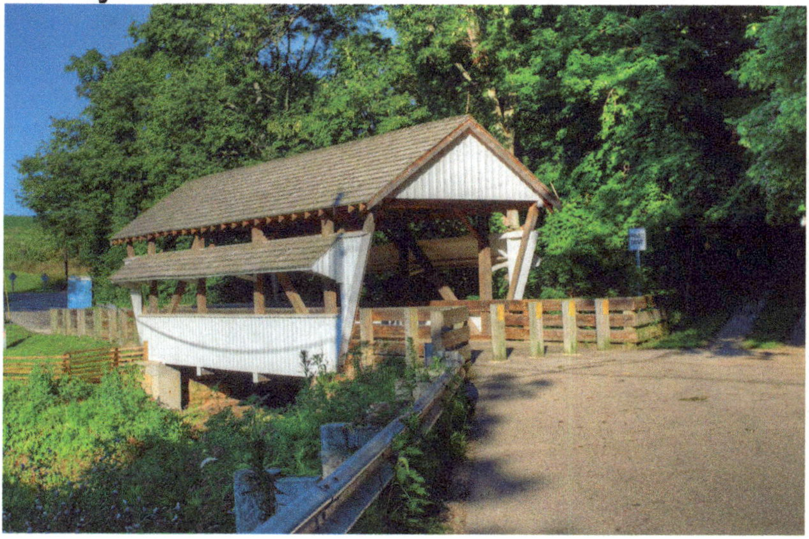

GPS Position: N 39° 44.963' W 82° 42.772'
Directions: From Wesley Chapel, head north on Rockmill Rd NW for 1.1 mi and turn right onto Lithopolis Rd NW. After 0.2 mi turn right onto Patton Ln where you will find the bridge
Crosses: Hocking River
Carries: Patton Lane, bypassed section of Rock Mill Road.
Builder: Jacob R. Brandt
Year Built: 1901 (R2003)
Truss Type: Queen
Dimensions: 1 Span, 37 feet

Notes: The bridge was closed and renovated in 2003. The 1824 Mill lies by a steep gorge beside it and includes a watermill. The whole site is a treat to visit.

World Index Number: OH/35-23-48
National Register of Historic Places: April 26, 1976

Dietz (Bergstresser) Covered Bridge

Township: Madison
County: Franklin

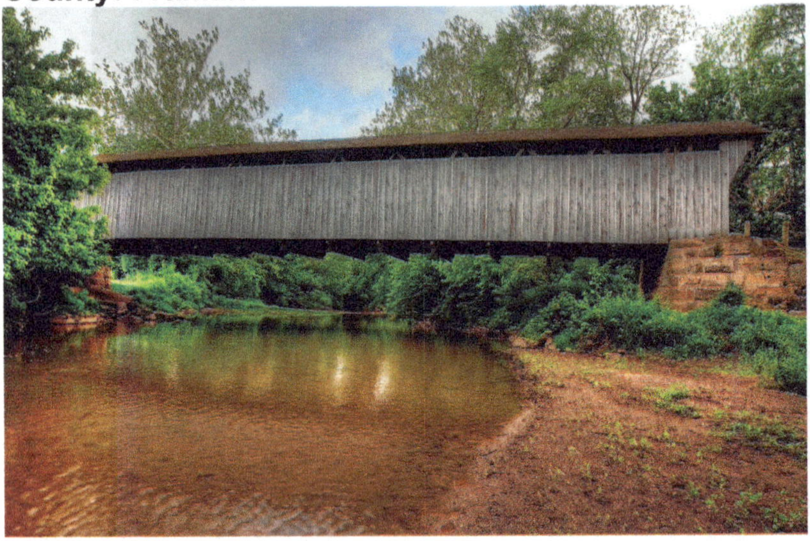

GPS Position: N 39° 49.790' W 82° 48.980'
Directions: In the town of Canal Westchester found on Ashbrook Road just north of Washington St.
Crosses: Walnut Creek
Carries: Ashbrook Road
Builder: R.L. Partridge, Columbus Bridge Co
Year Built: 1887 (R1971) (R1991)
Truss Type: Partridge
Dimensions: 1 Span, 125 feet
Notes: Reuban Partridge who constructed bridges in Union County, was the Vice-president of the Columbus Bridge Co. and he supervised the construction of this bridge using his patented truss design. After the 1991 renovation by the W.J. Seidensticker Company, ownership if the bridge was passed to the Village of Canal Winchester.
World Index Number: OH/35-25-03
National Register of Historic Places: May 3, 1974

Tare Creek Covered Bridge
Township: Middlefield
County: Geauga

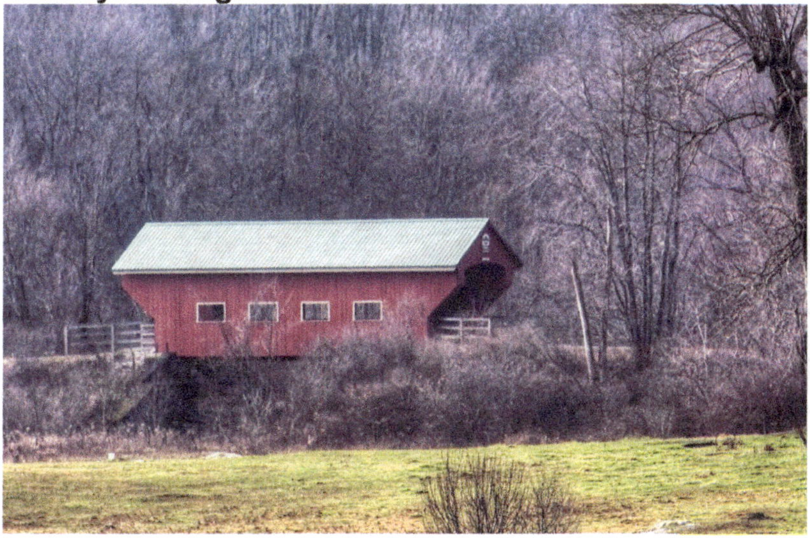

GPS Position: N 41° 28.831' W 81° 05.585'
Directions: From Middlefield go north on OH-608 for 1.9 mi and turn left on Burton-Windsor Rd. After 0.8 mi you will see the entrance to the Rail Trail. Go south on the trail for 0.7 miles where you will find the bridge.
Crosses: Tare Creek
Carries: Rail Trail
Builder: Geauga Parks Board
Year Built: 2004
Truss Type: Howe
Dimensions: 1 Span, 50 feet
Notes: The trail entrance can be a bit difficult to find. There is also access from Tare Creek Road found off OH-608/N State Ave., in the north of Middlefield. This recent addition to Ohio's bridges is a great looking structure.
World Index Number: OH/35-28-02#2
National Register of Historic Places: Not listed

Maple Highlands Trail Covered Bridge
Township: Middlefield
County: Geauga

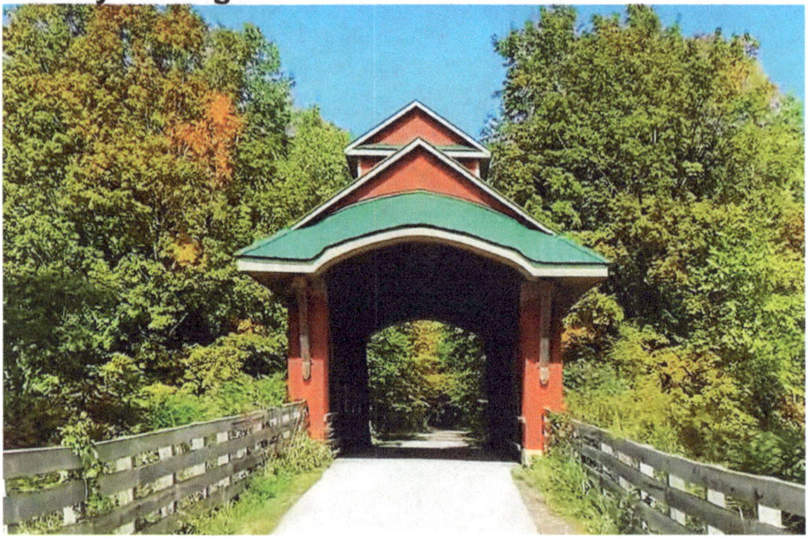

GPS Position: 41°29'41.0"N 81°06'11.3"W
Directions: From Burton, head east on Goodwin St for 0.5 mi and continue onto Burton-Windsor Rd. In 2.1 mi turn left onto Station Rd and the trail head is 0.3 mi. Walk 0.4 mi on the trail
Crosses: East Branch of the Cuyahoga River
Carries: Maple Highlands Trail
Builder: Smolen Engineering
Year Built: 2013
Truss Type: Pratt
Dimensions: 1 span 60 feet
Notes: Construction for this new bridge began in April, 2012 and was complete in August of 2013. The project cost about $180,000

World Index Number: OH/35-28-04
National Register of Historic Places: Not listed

Cemetery Road (Glen Helen, Yellow Springs) CB

Township: Miami
County: Greene

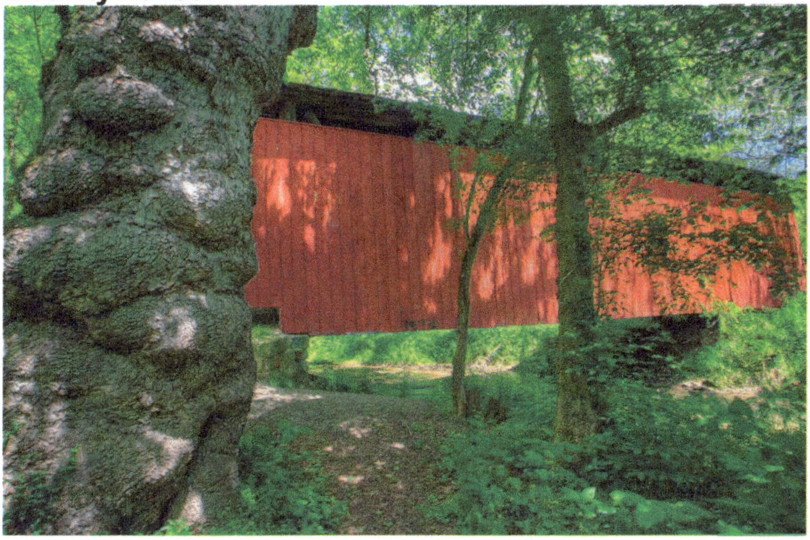

GPS Position: N 39° 47.130' W 83° 52.611'
Directions: From Yellow Springs take Corry St south off US-66 and after 0.9 mi turn left on Grinnell Rd. After passing the Grinnell Inn look for a pull off on the north side where a trail starts to the bridge.
Crosses: Yellow Springs Creek
Carries: Pedestrian path
Builder: Henry Hebble
Year Built: 1886 (M1979) (R2000)
Truss Type: Howe
Dimensions: 1 Span, 60 feet
Notes: The bridge was originally 130 feet long and spanned Anderson Creek. The center section was moved to its present location in 1979.

World Index Number: OH/35-29-01
National Register of Historic Places: Not listed

West Engle Mill Road Covered Bridge
Township: Caesars Creek
County: Greene

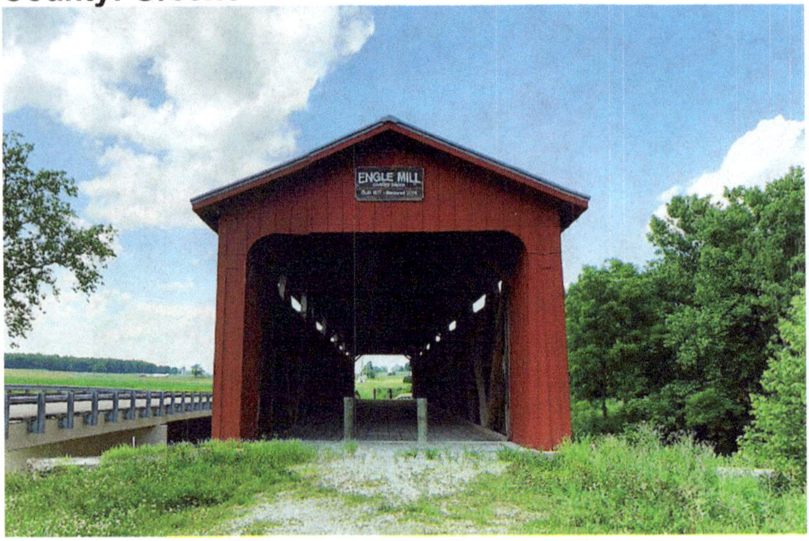

GPS Position: N 39°34.921' W 83° 55.322'
Directions: From the town of Spring Valley go east on Spring Valley/Paintersville Rd for 4.3 miles and turn right on Engle Mill Rd. where you will find the bridge after 0.7 miles on a bypassed section.
Crosses: Anderson's Fork
Carries: Engle Mill Rd.(Bypassed Section)
Builder: Smith Bridge Co.
Year Built: CA.1887 (R CA. 2015)
Truss Type: Smith Double
Dimensions: 1 Span, 136 feet
Notes: The bridge was bypassed in the 1980s when a new bridge was built nearby. The site looks great after the recent renovation.

World Index Number: OH/35-29-03
National Register of Historic Places: Not listed

Stevenson Road Covered Bridge
Township: Xenia
County: Greene

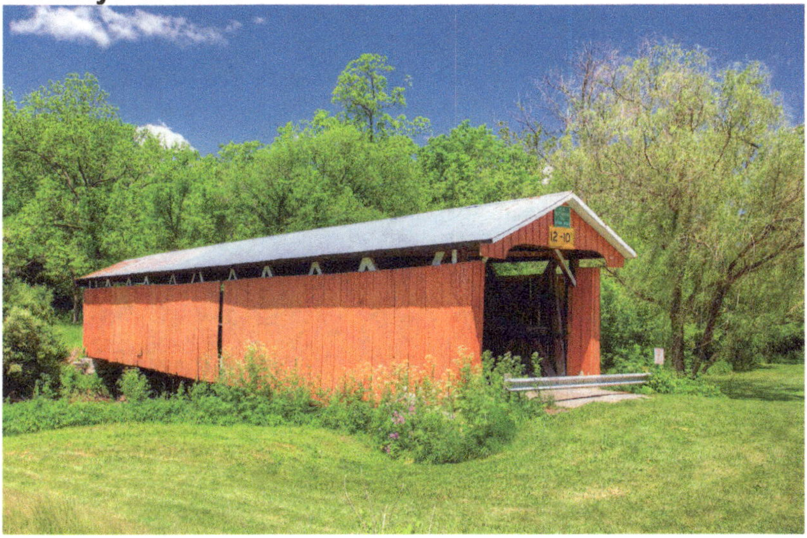

GPS Position: N 39° 44.104' W 83° 53.919'
Directions: From Xenia, go NW on US-42 for 0.7 mi and turn left on Stevenson Rd. where you will find the bridge in 2.3 mi.
Crosses: Massies Creek
Carries: Stevenson Road (Bypassed)

Builder: Smith Bridge Co
Year Built: 1877 (R1975)
Truss Type: Smith Double
Dimensions: 1 Span, 95 feet

Notes: The bridge was closed in 2003. The grant to rehabilitate it became controversial with politicians looking for a cause. It looks in great shape after that renovation.

World Index Number: OH/35-29-15
National Register of Historic Places: Not listed

Charleton Mill Road Covered Bridge

Township: Xenia
County: Greene

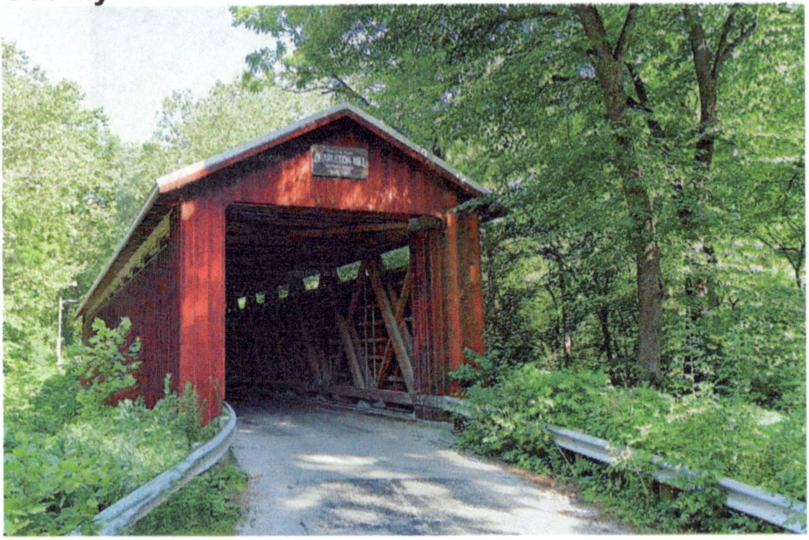

GPS Position: N 39° 43.737' W 83° 52.217'
Directions: Go east on Charlton Mill Rd. off US-42 southwest of the town of Cedarville.

Crosses: Massies Creek
Carries: Charlton Mill Rd.

Builder: Not known
Year Built: 2013
Truss Type: Howe
Dimensions: 1 Span, 120 feet

Notes: The bridge replaced an 1883 structure. It looks excellent

World Index Number: OH/35-29-16#2
National Register of Historic Places: Not listed

Ballard Road (Dean Farm) Covered Bridge

Township: New Jasper
County: Greene

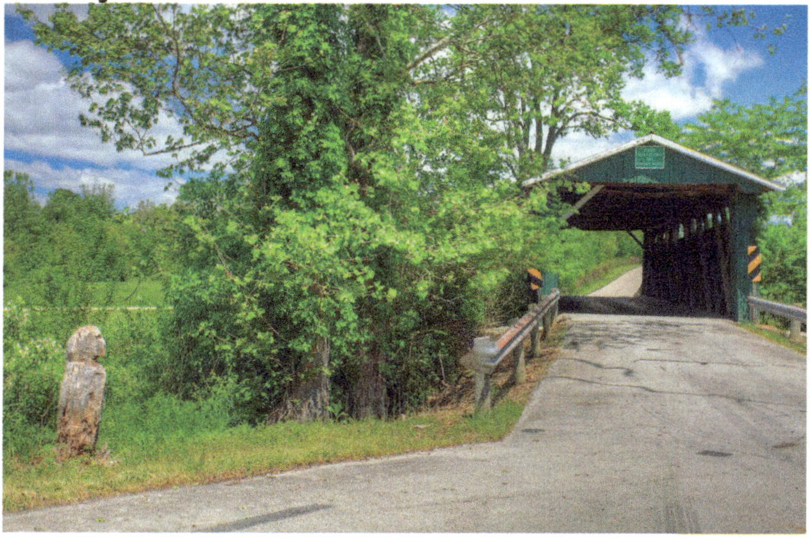

GPS Position: N 39° 40.674' W 83° 48.880'
Directions: From Cedarville go southeast on OH-72 for 2.0 miles and turn right on Federal Rd. Proceed for 4.3 miles and turn left on Old US-35 and after another 2.3 miles turn right on Ballard Road where you will find the bridge in 0.5 miles.
Crosses: Caesar Creek, North Branch
Carries: Ballard Road (Bypassed Section)
Builder: James E. Brown
Year Built: 1883
Truss Type: Howe
Dimensions: 1 Span, 74 feet

Notes: The bridge is closed and rests in a quiet rural setting. It was named for Lyman Ballard who had built a mill nearby.

World Index Number: OH/35-29-18
National Register of Historic Places: May 29, 1975

Richard P Eastman Hyde Road CB

Township: Miami
County: Greene

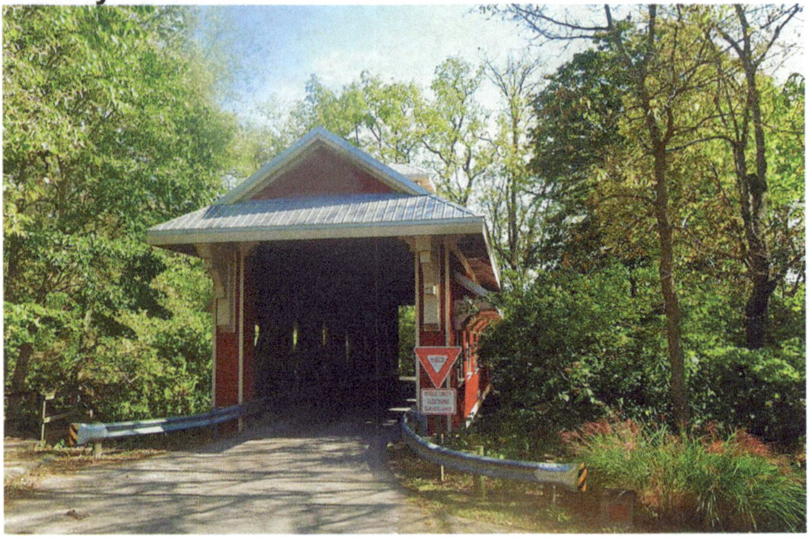

GPS Position: 39°47'06.8"N 83°53'11.8"W
Directions: 1030 E Hyde Rd, Yellow Springs
Crosses: East Hyde Road
Carries: Little Miami Bike Trail

Builder: R.G. Zachrich Construction
Year Built: 2014
Truss Type: Howe
Dimensions: 1 Span, 77 feet

Notes: The name honors Richard Eastman who was a county engineer who worked to save covered bridges

World Index Number: OH/35-29-109
National Register of Historic Places: Not listed

Indian Camp Covered Bridge

Township: Knox
County: Guernsey

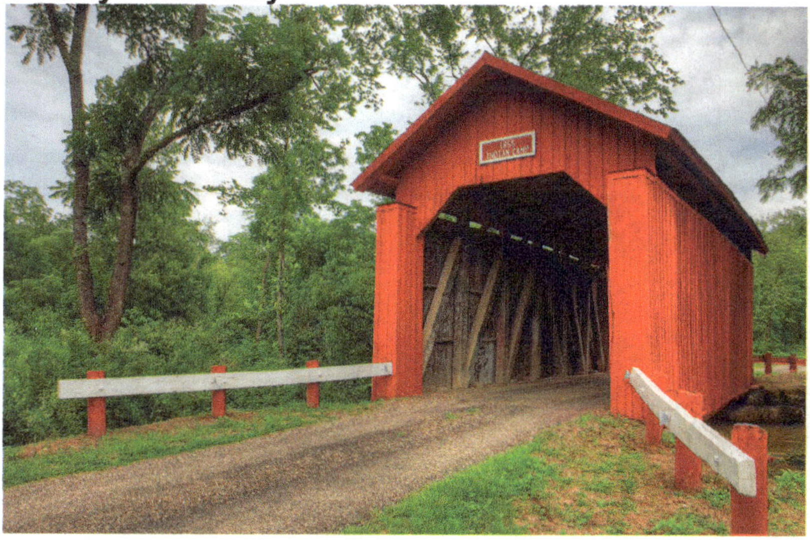

GPS Position: N 40° 06.775' W 81° 38.958'
Directions: From Cambridge go NW on OH-209/Bloomfield Rd for 1.6 mi and turn right on OH-658/Hopewell Rd. After 5.6 mi turn right on Covered Bridge Rd and see the bridge.
Crosses: Indian Camp Creek
Carries: Covered Bridge Road
Builder: Not known.
Year Built: 1855 (R1963) (R1999)
Truss Type: Multiple Kingpost
Dimensions: 1 Span, 36 feet
Notes: The bridge deck was damaged in 1998 due to flooding and the bridge was closed. It was re-opened in 1999 after repairs were completed. At some point it was reinforced with steel beams.

World Index Number: OH/35-30-04
National Register of Historic Places: Not listed

Armstrong (Clio) Covered Bridge

Township: Cambridge
County: Guernsey

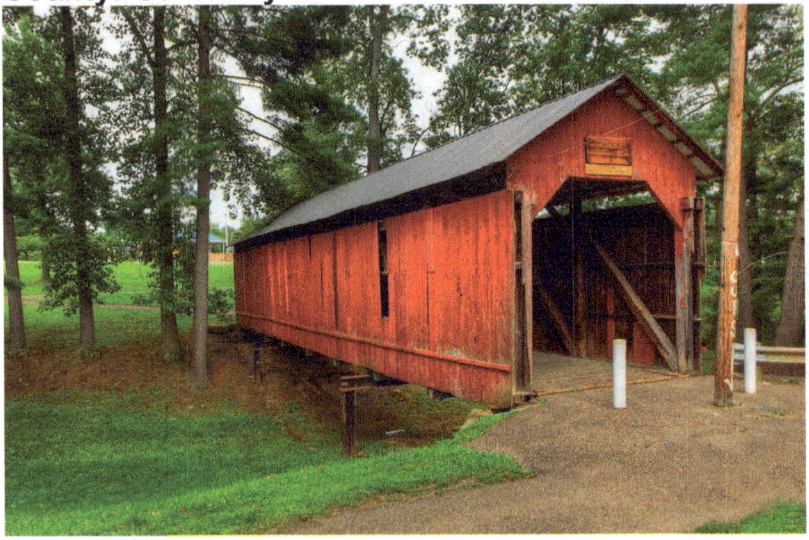

GPS Position: N 40° 02.151' W 81° 35.387'
Directions: By McKinley Road east of 8th St. in the town of Cambridge.
Crosses: None
Carries: None
Builder: Abraham Armstrong
Year Built: 1849 (M1966)
Truss Type: Multiple Kingpost
Dimensions: 1 Span, 76 feet
Notes: One of Ohio's oldest bridges it was originally located at Clio, Ohio, but was relocated when a reservoir was constructed which would have flooded the site. It was the only one of three bridges at Clio which were saved. Abraham Armstrong constructed the bridge to provide access to his mill.

World Index Number: OH/35-30-12
National Register of Historic Places: Not listed

Jediah Hill (Groff Mill) Covered Bridge

Township: Springfield
County: Hamilton

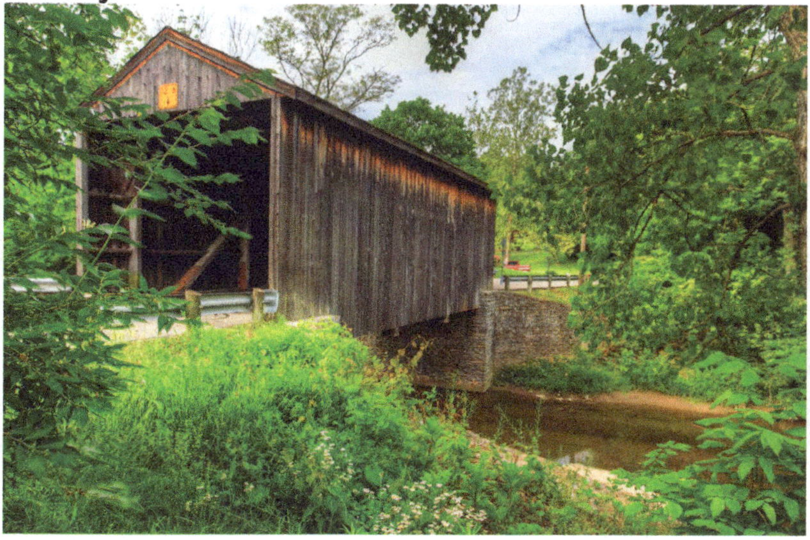

GPS Position: N 39° 15.154' W 84° 32.706'
Directions: In Cincinnati, take exit 36 off I-275 and go south on US-127 for 2.6 mi and turn left on Miles Rd. After 0.1 miles turn right on Covered Bridge Rd where the bridge is a short distance.
Crosses: Mill Creek, West Fork
Carries: Covered Bridge Road
Builder: Jediah Hill
Year Built: 1850 (R1956) (R1981)
Truss Type: Queen
Dimensions: 1 Span, 44 feet
Notes: Still open to traffic, it was built by Jediah Hill for access to his mill. The 1956 repairs were necessitated when the bridge was damaged by an ice storm.

World Index Number: OH/35-31-01
National Register of Historic Places: March 28, 1973

Blanchard River Covered Bridge

Township: Marion
County: Hancock County

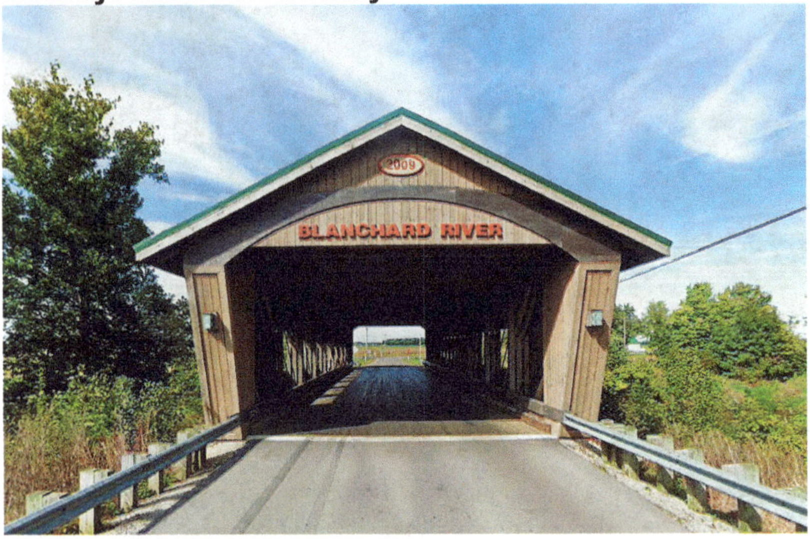

GPS Position: 41°02'06.2"N 83°33'50.6"W
Directions: From Findlay, head east on E Sandusky St for 1.9 mi and continue onto OH-568 E/Carey Rd for 2.5 mi. Turn right onto Trail 241 and the bridge
Crosses: Blanchard River
Carries: Trail 241

Builder: Not known
Year Built: 2009
Truss Type: Pratt
Dimensions: 1 Span, 125 feet

Notes: This new bridge is easy and open to photograph from all sides. The long side windows are a nice detail

World Index Number: OH/35-32-21
National Register of Historic Places: Not listed

Skull Fork Covered Bridge
Township: Freeport
County: Harrison

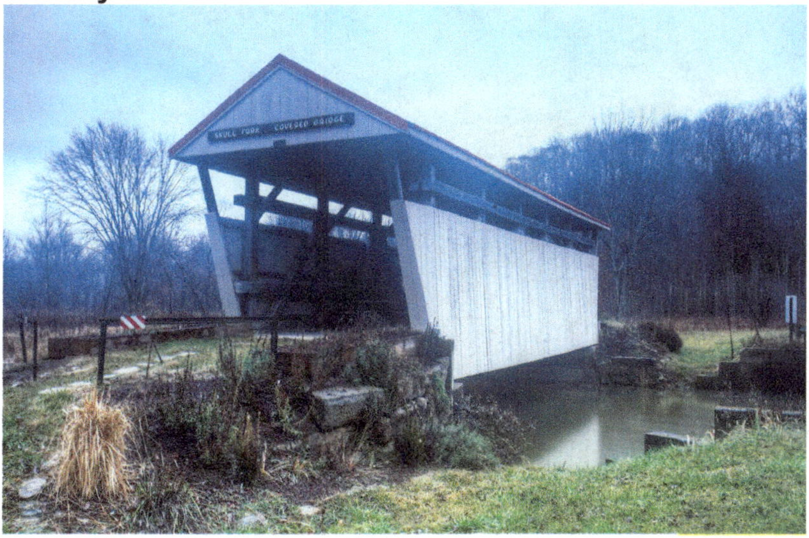

GPS Position: N 40° 10.976' W 81° 16.153'
Directions: From Freeport go south on Skull Fork Rd from OH-800 for 1.9 miles where you will see the bridge.
Crosses: Skull Fork
Carries: Skull Fork Rd

Builder: Not known
Year Built: 1876 (R2006)
Truss Type: Multiple Kingpost
Dimensions: 1 Span, 45 feet

Notes: This bridge is bypassed by a modern concrete bridge. The 2006 renovation has it looking in great condition and there is a small park is at the site.

World Index Number: OH/35-34-19
National Register of Historic Places: Not listed

Stutzman's Crossing Covered Bridge
Township: Walnut Creek
County: Holmes

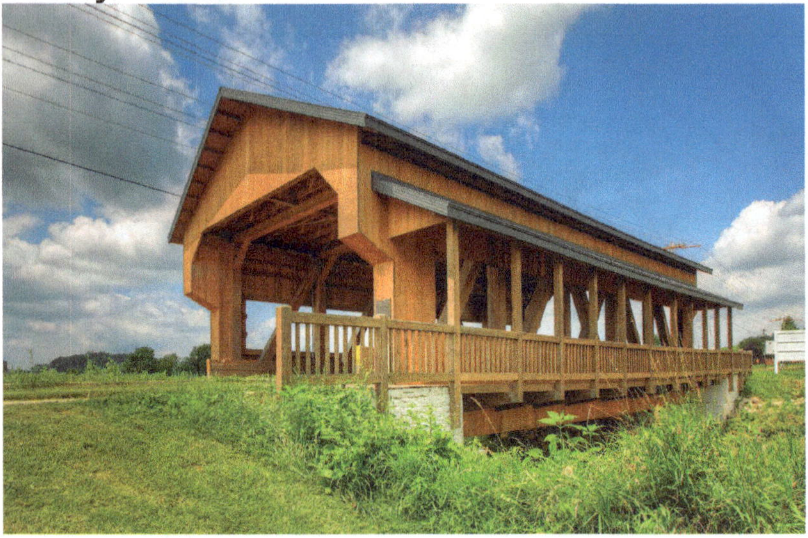

GPS Position: N 40° 31.906' W 81° 43.105'
Directions: From the town of Sugarcreek, go west on OH-39 for 3.5 miles and turn right on CR-114 and after 400 feet, turn right on CR-145 where you will find the bridge.
Crosses: Walnut Creek
Carries: CR-145
Builder: Jonas Erb Builders.
Year Built: 2009
Truss Type: Multiple Kingpost
Dimensions: 1 Span, 56 feet
.
Notes: Another of Ohio's many new covered bridges, it is a great looking structure which incorporates a pedestrian walkway as well.

World Index Number: OH/35-38-04
National Register of Historic Places: Not listed

Spellacy Covered Bridge
Township: Knox
County: Holmes

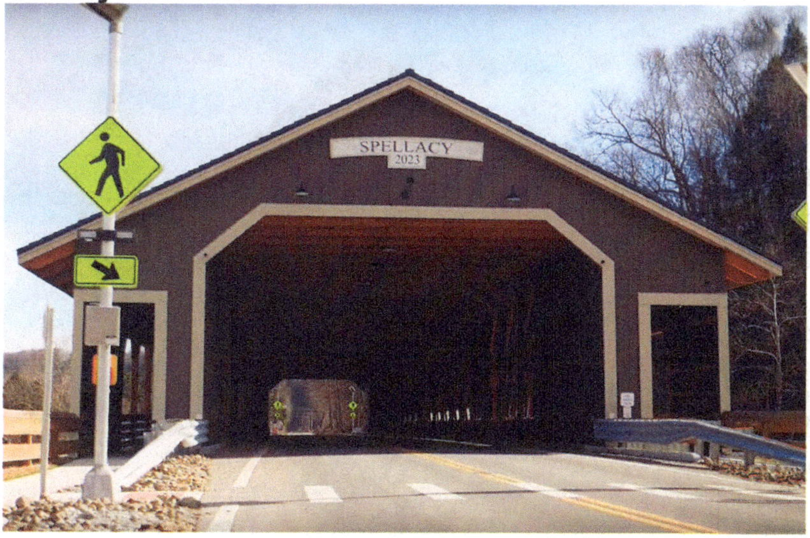

GPS Position: 40°35'38.7"N 82°12'25.1"W
Directions: From Loudonville, head south on S Market St for 0.5 mi and turn left onto Co Rd 529. After 2.2 mi continue onto Co Hwy 230/County Rd 230 for 0.7 mi and turn left onto Co Hwy 23 and the site is 0.3 mi
Crosses: Mohican River
Carries: Co Hwy 23
Builder: Kokosing Construction
Year Built: 2023
Truss Type: Modified Pratt
Dimensions: 2 Span, 300 feet
.
Notes: A great looking new structure which includes 2 lanes and 2 pedestrian walkways as well.

World Index Number: OH/35-38-05
National Register of Historic Places: Not listed

Johnson Road (Crabtree) Covered Bridge

Township: Scioto
County: Jackson

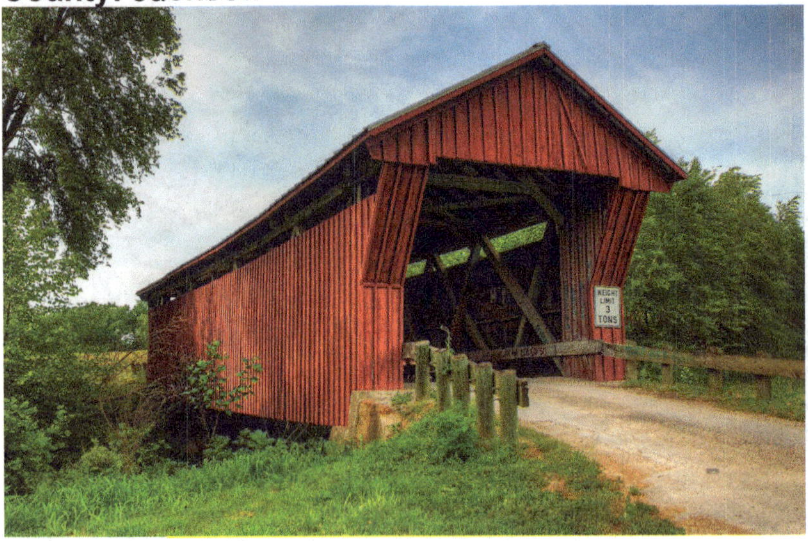

GPS Position: N 38° 57.502' W 82° 47.261'
Directions: From Jackson, head west on OH-776 and after 8.3 mi turn left on Spangenburg Rd. In 0.9 mi make a slight right onto Johnson Rd where you will find the bridge in 1.2 mi.
Crosses: Little Scioto River
Carries: Johnson Road
Builder: Not known
Year Built: 1870 (R2000)
Truss Type: Smith
Dimensions: 1 Span, 71 feet

Notes: It had been closed to traffic for many years, but was re-opened after the 2000 renovation. It looks in excellent shape although the graffiti vandals have been at the interior.

World Index Number: OH/35-40-06
National Register of Historic Places: August 23, 1984

Byer Covered Bridge
Township: Washington
County: Jackson

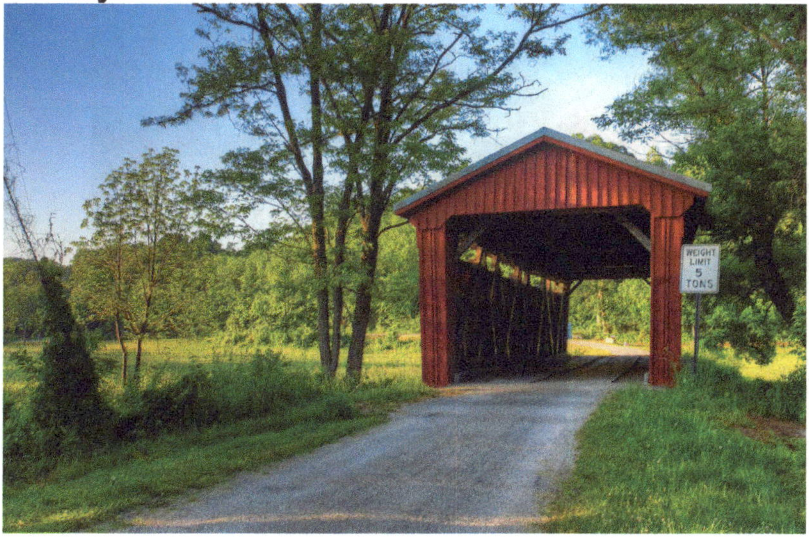

GPS Position: N 39° 10.791' W 82° 37.883'
Directions: In the town of Byer found by Byers Winter Road just south of Main St.
Crosses: Pigeon Creek
Carries: Byers Winter Road

Builder: T.J. Dency
Year Built: 1872 (R1935) (R2003)
Truss Type: Smith Double
Dimensions: 1 Span, 74 feet

Notes: Re-opened to traffic after the 2003 renovation. It had been closed for many years and the original stone abutments have had a concrete casing added.

World Index Number: OH/35-40-08
National Register of Historic Places: October 21, 1975

Buckeye Furnace Covered Bridge
Township: Milton
County: Jackson

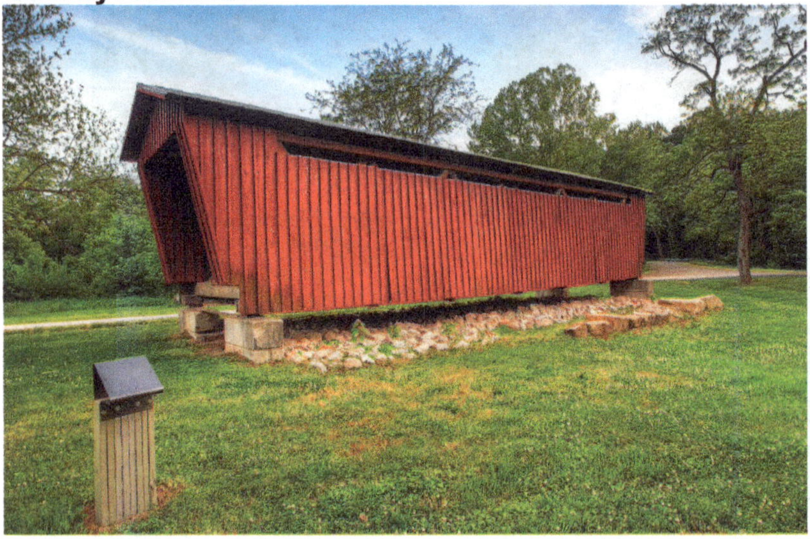

GPS Position: N 39° 03.268' W 82° 27.518'
Directions: From Wellston, go south on OH-327 for 1.7 mi and turn left on OH-124. After 4.0 mi turn right on CR-58/Buckeye Furnace Rd. In 2.3 mi turn right on Buckeye Park Road and proceed for 0.7 miles where you will find the bridge.
Crosses: Dry ground
Carries: Buckeye Park Road (Bypassed)

Builder: Dency, McCurdy & Company
Year Built: 1871 (R1867) (R1936) (R1960s) (R1999) (M2011)
Truss Type: Smith Double
Dimensions: 1 Span, 59 feet
Notes: The bridge was moved across the road when a new bridge replaced it in 2011.

World Index Number: OH/35-40-11
National Register of Historic Places: February 24, 1975

Scottown (Pleasant Ridge) Covered Bridge
Township: Windsor
County: Lawrence

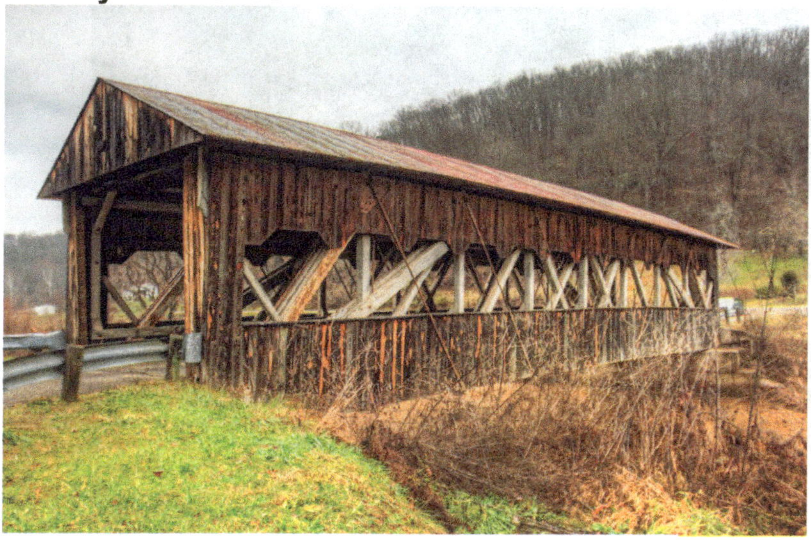

GPS Position: N 38° 32.887' W 82° 22.814'
Directions: From Scottown go east on OH-217 for 0.6 mi and turn right on Pleasant Ridge Rd. where you will see the bridge.
Crosses: Indian Guyan Creek
Carries: Pleasant Ridge Rd. (Bypassed)
Builder: William Thompson
Year Built: 1877 (R1913) (R1934) (R1971) (R1991)
Truss Type: Multiple Kingpost (Burr Arch Added)
Dimensions: 1 Span, 79 feet
Notes: This structure survived a couple of early floods. In 1934 a modified Burr Arch and steel tie rods were added after a flood known as The Great Wheat Flood. Fields of wheat washed into the creek, then hit the bridge and almost knocked it off its abutments. In 1971 the roof was replaced.

World Index Number: OH/35-44-05
National Register of Historic Places: November 12, 1975

Boy Scout (Rainrock) Covered Bridge

Township: Eden
County: Licking

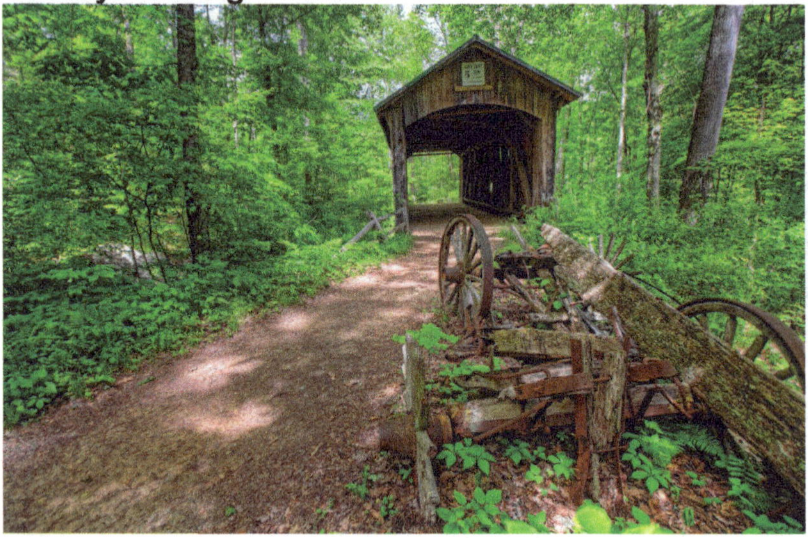

GPS Position: N 40° 10.587' W 82° 18.355'
Directions: From Newark go north on OH-79 for 8.0 mi and turn left on Rock Fork Rd. After 0.6 mi turn left on Houdeshell Rd. and the entrance to Camp Falling Rock is 0.7 mi. The bridge is on a interior road but permission is required to visit.
Crosses: Rocky Fork
Carries: Interior camp road
Builder: Not known
Year Built: Not known (M1974)
Truss Type: Multiple Kingpost
Dimensions: 1 Span, 49 feet

Notes: Originally located a mile south of its current location, in 1976, the bridge was moved to its the present location.

World Index Number: OH/35-44-04
National Register of Historic Places: Not listed

Girl Scout (Shoults) Covered Bridge
Township: Fallsbury
County: Licking

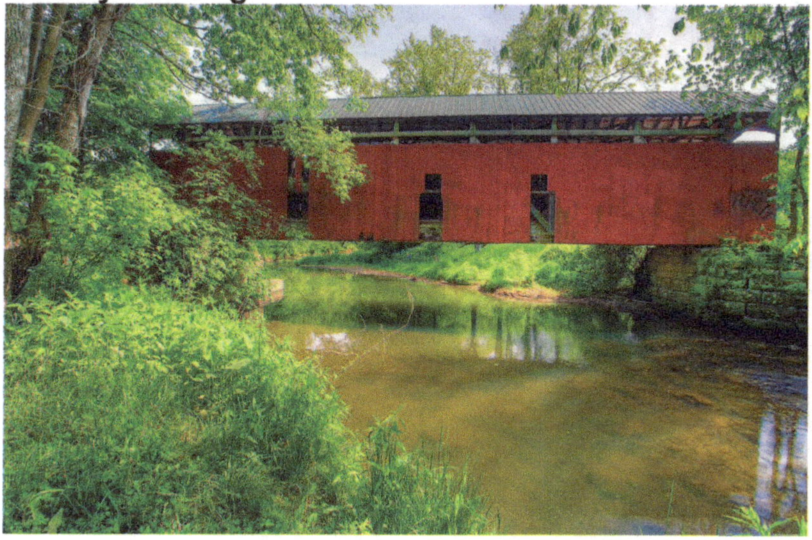

GPS Position: N 40° 14.006' W 82° 14.974'
Directions: From the town of Fallsburg, go north on OH-586 for 2.4 miles and turn right on Girl Scout Camp Road where you will find the bridge in 1.0 miles.
Crosses: Wakatomika Creek
Carries: Girl Scout Camp Road
Builder: Not known
Year Built: 1879 (R2003) (R2008)
Truss Type: Multiple Kingpost
Dimensions: 1 Span, 68 feet
Notes: The bridge is not in good shape with extensive graffiti and siding missing. It is at a favorite swimming hole and often has side boards knocked out so that the swimmers can jump in.

World Index Number: OH/35-45-05
National Register of Historic Places: Not listed

Gregg Mill (Pine Bluff) Covered Bridge

Township: Fallsbury
County: Licking

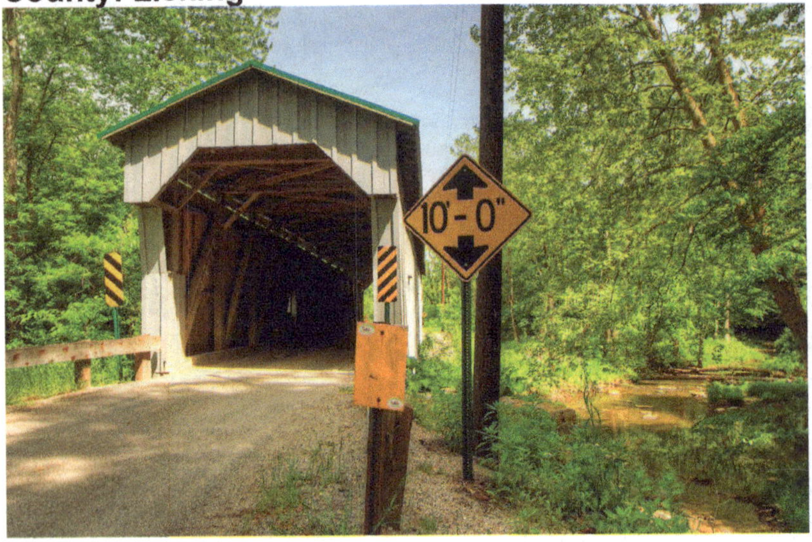

GPS Position: N 40° 13.315' W 82° 14.333'
Directions: From Fallsburg, go north on OH-586 for 1.2 mi and turn right on Frampton Rd. where you will find the bridge after 1.2 miles.
Crosses: Wakatomika Creek
Carries: Frampton Rd.
Builder: Not known
Year Built: 1993 (R1993)
Truss Type: Multiple Kingpost
Dimensions: 1 plus Span, 124 feet

Notes: This bridge built in 1993 replacing an 1881 bridge. It is in a quiet setting. It has two windows on the downstream side and three on the upstream.

World Index Number: OH/35-45-06#2
National Register of Historic Places: Not listed

Lobdell Park (McLain) Covered Bridge
Township: St. Albans
County: Licking

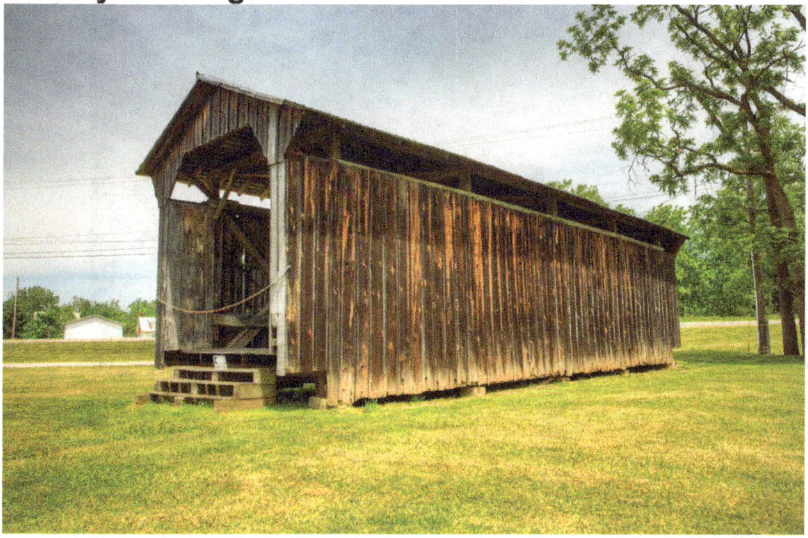

GPS Position: N 40° 05.187' W 82° 36.483'
Directions: In the town of Alexandria found in Fireman's Park by Main St/OH-37 south of the intersection with Granville St.
Crosses: None
Carries: None
Builder: Not known
Year Built: 1871 (R1940s) (R1977) (M1977)
Truss Type: Multiple Kingpost
Dimensions: 1 Span, 47 feet

Notes: The bridge was originally situated at the intersection of TR 116 and Lobdell Road, crossing Lobdell Creek. It was bypassed in 1970 and abandoned until 1977 when it was moved to its present location and renovated.

World Index Number: OH/35-45-17
National Register of Historic Places: Not listed

Davis Farm Covered Bridge
Township: Maryann
County: Licking

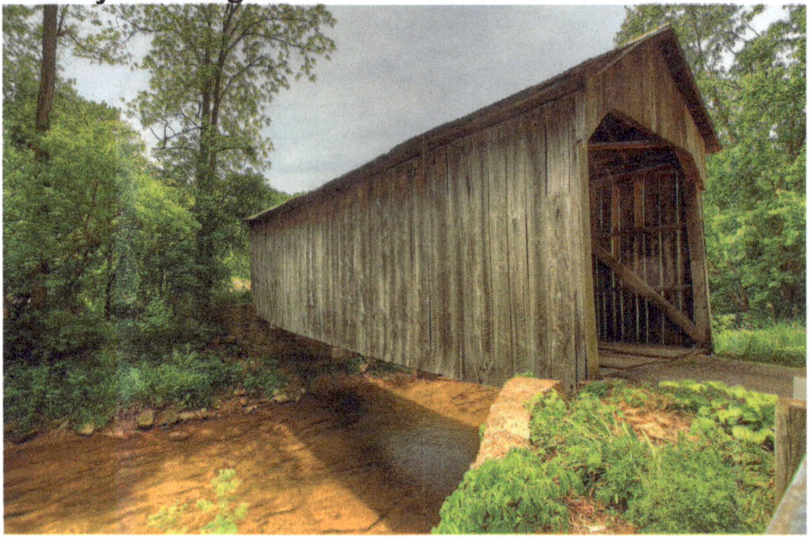

GPS Position: N 40° 08.099' W 82° 18.136'
Directions: From Newark, go north on OH-79 for 6.6 mi and turn right on Hickman Rd. Continue on Hickman Rd for 1.1 miles where you will see the bridge by the side of the road.
Crosses: Rocky Fork
Carries: Private road off Hickman Road

Builder: Not known
Year Built: 1947
Truss Type: Multiple Kingpost
Dimensions: 1 Span, 50 feet

Notes: On private property, the bridge was constructed from barn timbers.

World Index Number: OH/35-45-25
National Register of Historic Places: Not listed

McColly Covered Bridge
Township: Bloomfield-Washington
County: Logan

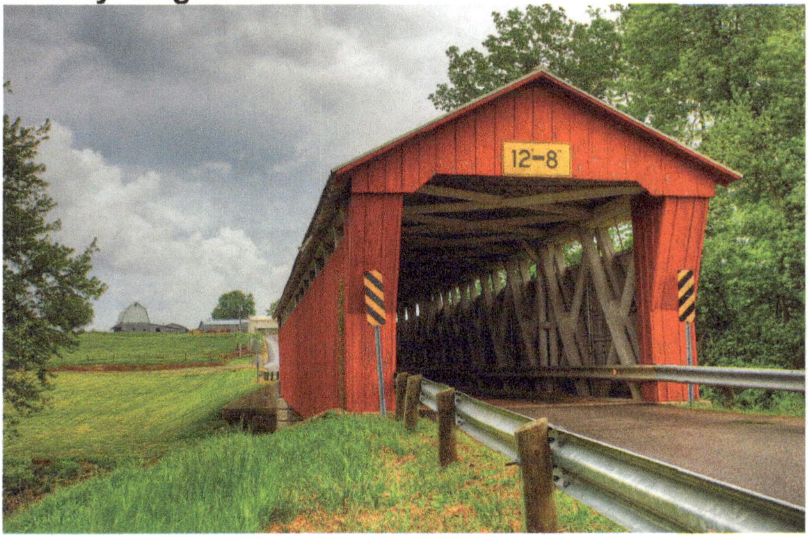

GPS Position: N 40° 24.078' W 83° 55.536'
Directions: From Huntsville go west on OH-274 for 4.7 miles and turn left on OH-235 and proceed for 3.1 miles. Turn right on CR-13 and you will find the bridge after 1.1 miles.
Crosses: Great Miami River
Carries: County Rd 13
Builder: Anderson Green Co
Year Built: 1876 (R1943) (R1958) (R2000)
Truss Type: Howe
Dimensions: 1 Span, 125 feet

Notes: The last restoration by Amos B. Schwartz Construction Company which brought the bridge from imminent collapse, has left it in beautiful condition in a quiet setting.

World Index Number: OH/35-46-01
National Register of Historic Places: 05/28/1975

Bickham Covered Bridge
Township: Richland
County: Logan

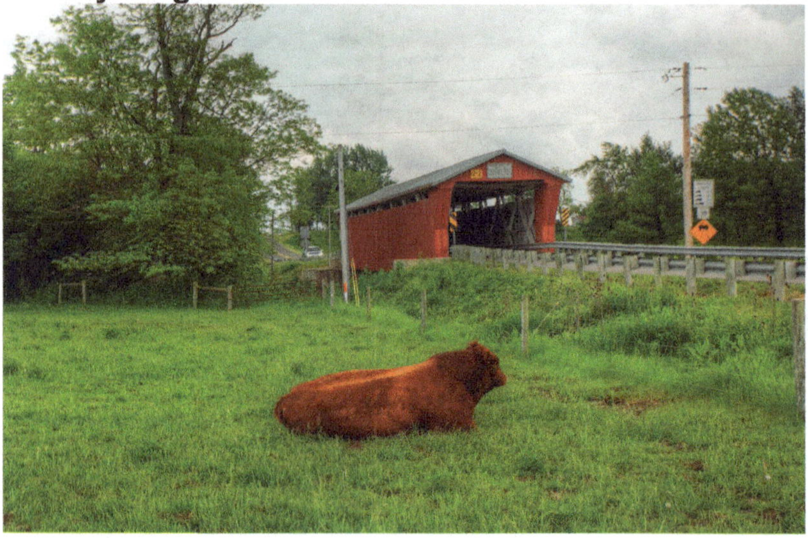

GPS Position: N 40° 28.494' W 83° 50.422'
Directions: From Russell's Point, go east on OH-366 for 1.2 mi and turn left on CR-38 and the bridge is 0.3 mi.
Crosses: Great Miami River, South Fork
Carries: CR-38

Builder: Smith Bridge Co. (R1959 Richter Construction)
Year Built: 1877 (R1959) (R2002)
Truss Type: Howe
Dimensions: 1 Span, 94 feet

Notes: The bridge is still active, It is in a quiet setting except for birdsong and occasional traffic. The 2002 renovation has the structure in beautiful shape.

World Index Number: OH/35-46-03
National Register of Historic Places: Not listed

Taylor Blair Covered Bridge
Township: West Jefferson
County: Madison

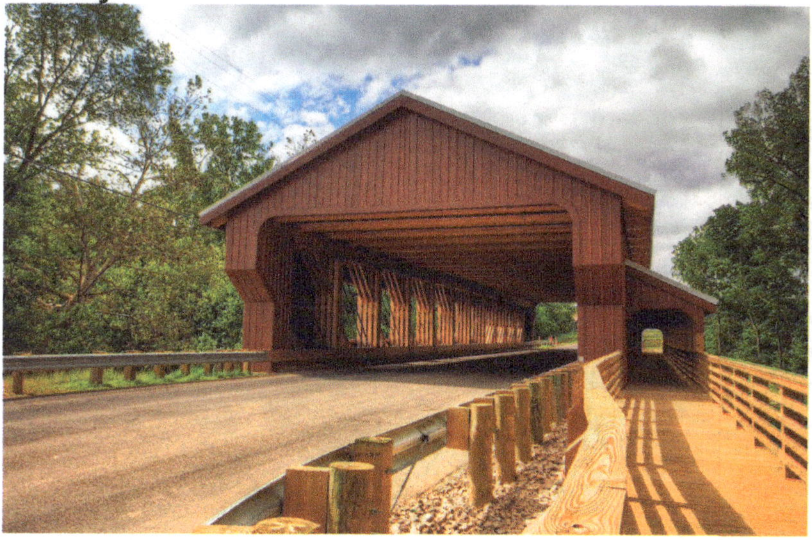

GPS Position: N 39° 56.816' W 83° 16.904'
Directions: In Columbus, go west on US-40 8.4 mi from the junction of I-270 and turn right on Taylor Blair Rd. where the bridge is a short distance.
Crosses: Little Darby Creek
Carries: Taylor Blair Rd.
Builder: R.B. Jergens
Year Built: 2012
Truss Type: Pratt
Dimensions: 1 Span, 150 feet
Notes: This is one of the latest covered bridges built in Ohio. It is a great looking structure with not only a wide double lane for high load traffic but a pedestrian walkway as well. Both the interior Pratt truss and the tamping system on the underside of the deck are interesting and worth examining.
World Index Number: OH/35-49-27
National Register of Historic Places: Not listed

St James Road Covered Bridge
Township: Richland
County: Marion

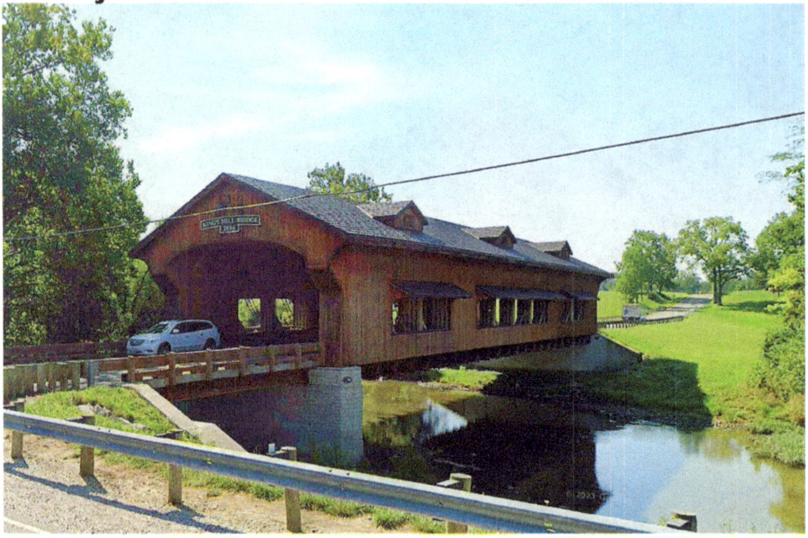

GPS Position: 40°30'44.5"N 83°03'11.9"W
Directions: From Waldo, head NE on OH-98 N for 3.5 mi and turn right onto Whetstone River Rd S. After 0.4 mi turn right onto St James Rd and the bridge
Crosses: Olentangy River
Carries: St James Road

Builder: RG Zachrich Construction
Year Built: 2016
Truss Type: Modified Pratt
Dimensions: 1 Span, 190 feet

Notes: This new structure is exceptional looking with its side windows. Easy setups from all sides

World Index Number: OH/OH/35-51-21
National Register of Historic Places: Not listed

Allen's Mill (Eldean) Covered Bridge
Township: Concord-Stauton
County: Miami

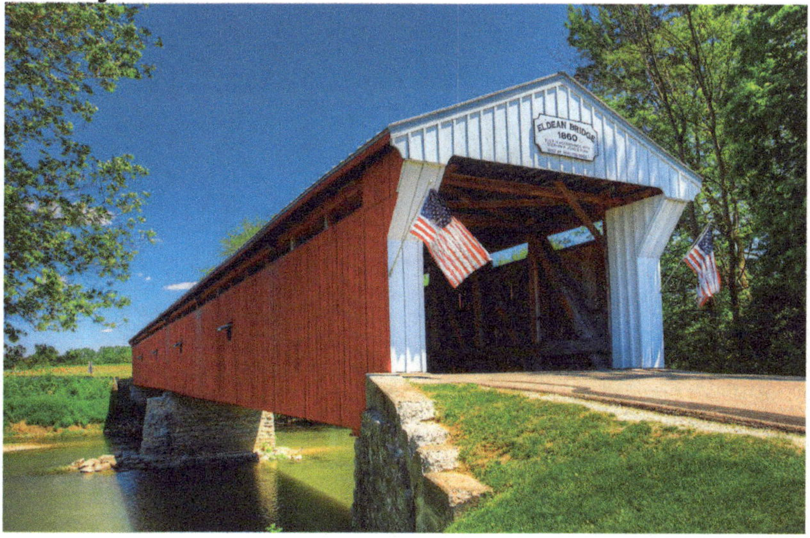

GPS Position: N 40° 04.690' W 84° 13.041'
Directions: From Piqua, go south on CR-25A/Main St for 3.8 mi and turn left on Eldean Rd. and see the bridge
Crosses: Great Miami River
Carries: Eldean Rd. (Bypassed section)

Builder: James and William Hamilton
Year Built: 1860 (R1980) (R2006)
Truss Type: Long
Dimensions: 2 Span, 224 feet

Notes: The bridge was bypassed in 1963 although it still carries some traffic as an entrance to a park. It is the longest bridge surviving which uses the Long truss system.

World Index Number: OH/35-55-01
National Register of Historic Places: February 20, 1975

Foraker (Weddle) Covered Bridge
Township: Perry
County: Monroe

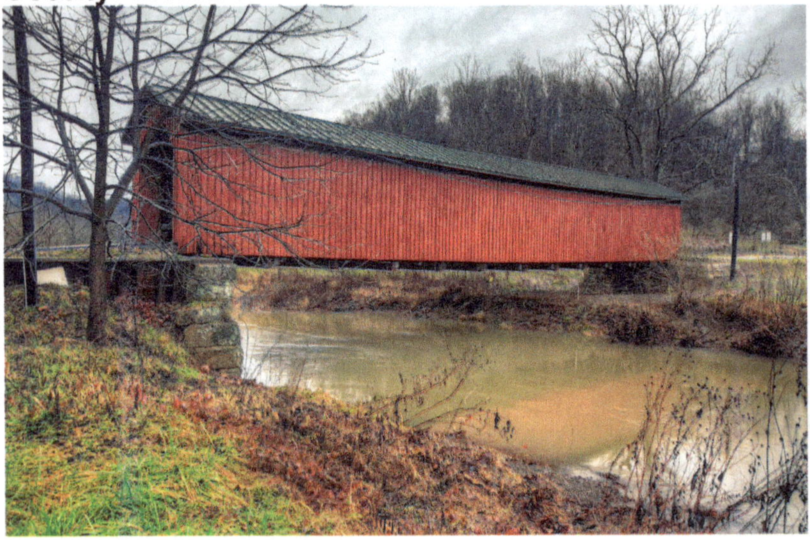

GPS Position: N 39° 39.364' W 81° 07.246'
Directions: From the town of Graysville go east on CR-12/Greenbrier Rd for 3.5 miles and turn left on Plainview Rd. The bridge is about 1 mile.
Crosses: Little Muskingum River
Carries: Plainview Rd.

Builder: Isiah Cline
Year Built: 1887 (R1938) (R2005)
Truss Type: Multiple Kingpost
Dimensions: 1 Span, 99 feet

Notes: The bridge was in poor shape until the renovation in 2005 which has it in great condition. It is still open to traffic.

World Index Number: OH/35-56-14
National Register of Historic Places: June 5, 1975

Long (Knowlton, Old Camp) Covered Bridge
Township: Washington
County: Monroe

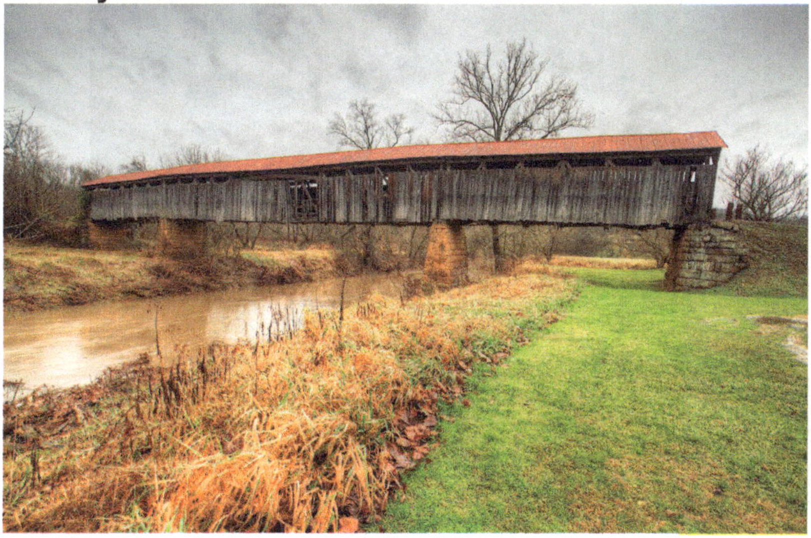

GPS Position: N 39° 36.089' W 81° 09.451'
Directions: From the village of Rinard Mills go north on OH-26/Hilight for 1.4 miles and turn right Township Rd 384. The bridge is at the end of the road.
Crosses: Little Muskingum River
Carries: Township Rd 384 (Bypassed)
Builder: Not known
Year Built: 1887 (R1896) (R1923) (R1928) (R1938) (R1996)
Truss Type: Multiple Kingpost and Burr Arch
Dimensions: 3 plus Span, 192 feet

Notes: This massive structure has been bypassed and is showing its age. Originally only the center span was covered with the outer spans being covered in 1884.

World Index Number: OH/35-56-18
National Register of Historic Places: March 11, 1980

Germantown Covered Bridge

Township: German
County: Montgomery

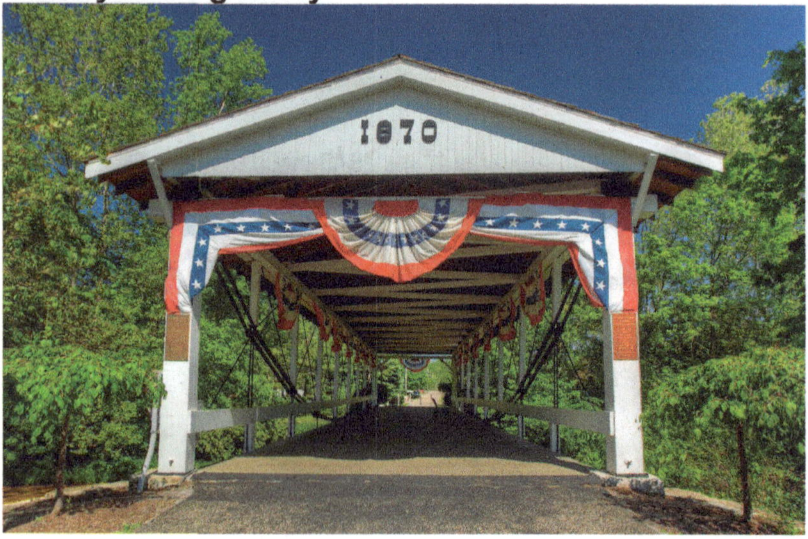

GPS Position: N 39° 37.571' W 84° 21.942'
Directions: Found in the town of Germantown on Center St. east of Water St.
Crosses: Little Twin Creek
Carries: Center Street (Bypassed)
Builder: David Morrison
Year Built: 1865 (R1906) (M1911) (R1963) (R1981)
Truss Type: Inverted Bowstring
Dimensions: 1 Span, 100 feet

Notes: This bridge originally spanned Little Twin Creek on the Dayton Pike but was moved to this location in 1911. There are only two bridges which use an Inverted Bowstring Truss and this is the only one that uses it as its sole truss.

World Index Number: OH/35-57-01
National Register of Historic Places: Not listed

Feedwire (Carillon Park) Covered Bridge
Township: Harrison
County: Montgomery

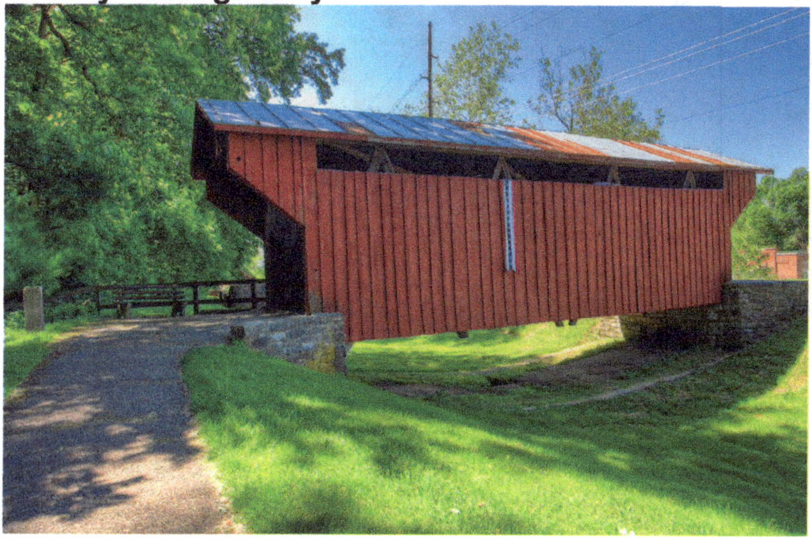

GPS Position: N 39° 43.794' W 84° 11.963'
Directions: In Dayton, in the Carillon Historical Park near the intersection of Carillon Blvd and Patterson Blvd. A fee is necessary if you wish to tour the historical buildings on site.
Crosses: Miami and Erie Canal Bed
Carries: N/A
Builder: Smith Bridge Company
Year Built: 1870 (M1948)
Truss Type: Smith Single and Burr Arch
Dimensions: 1 Span, 42 feet
Notes: Originally this bridge was built near Bellbrook to cross the Little Sugar Creek. It was moved to the Carillon Historical Park in 1948 and is part of their exhibit of historical buildings.

World Index Number: OH/35-57-03
National Register of Historic Places: Not listed

Jaspar Road (Mud Lick) Covered Bridge
Township: German
County: Montgomery

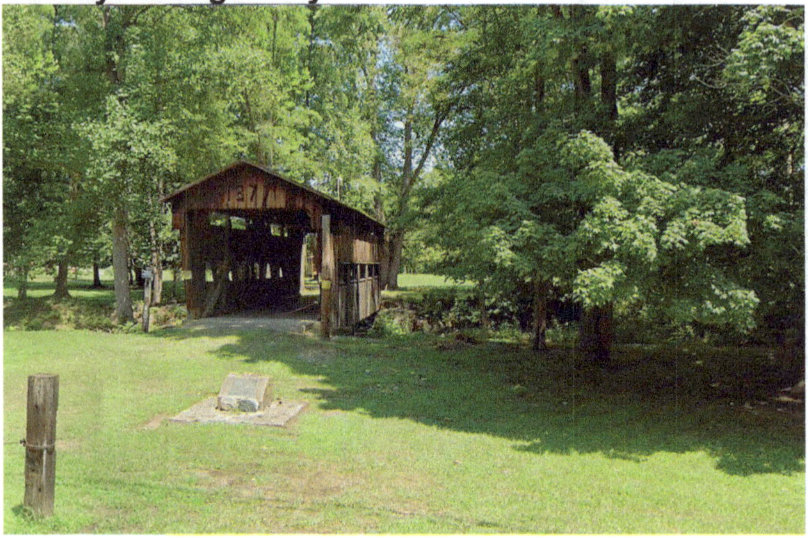

GPS Position: N 39° 37.036' W 84° 23.35'
Directions: From Germantown, go south on Astoria Rd. from OH-725 for 0.5 miles and turn right on Astoria Rd, continuing on Oxford Rd. The property with the bridge site is 0.4 miles.
Crosses: Mudlick Creek
Carries: Private road
Builder: Smith Bridge Co.
Year Built: 1869 (M1964)
Truss Type: Smith Single and Burr Arch
Dimensions: 1 Span, 50 feet

Notes: On private property but can be viewed from the road. Originally built in Greene County, OH and later moved to its present location.

World Index Number: OH/35-57-36
National Register of Historic Places: Not listed

Barkhurst Mill (Williams) Covered Bridge
Township: Marion
County: Morgan

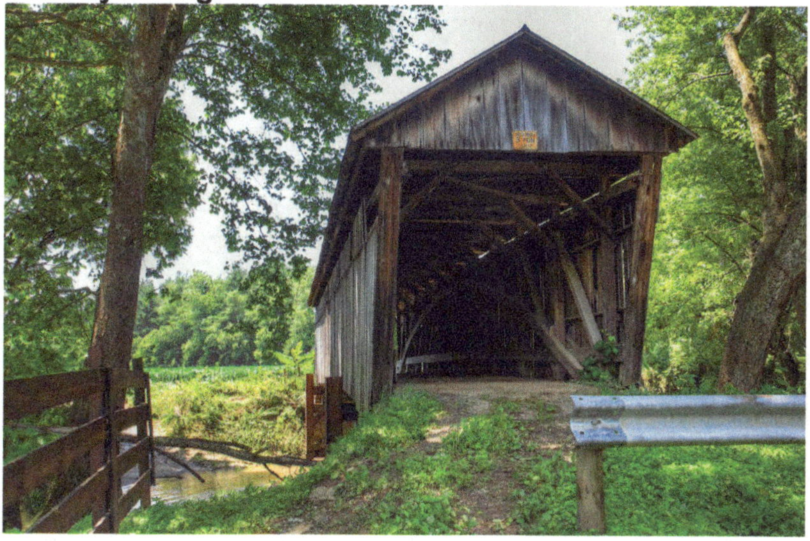

GPS Position: N 39° 30.156' W 81° 50.381'
Directions: From Chesterhill, go NE on OH-377 and after 0.9 miles turn right on CR-52/Goshen Run Rd, continuing onto Township Road 21/Williams Bridge Rd where you will find the bridge in 1.0 mile.
Crosses: Wolf Creek
Carries: Township Road 21/Williams Bridge Road (Bypassed)
Builder: Benjamin and John Pierpoint
Year Built: 1872
Truss Type: Multiple Kingpost and Burr Arch
Dimensions: 1 Span, 84 feet

Notes: The bridge lies abandoned at the end of Township Road 21.

World Index Number: OH/35-58-15
National Register of Historic Places: February 5, 1999

Rosseau (Fairgrounds) Covered Bridge
Township: Rosseau
County: Morgan

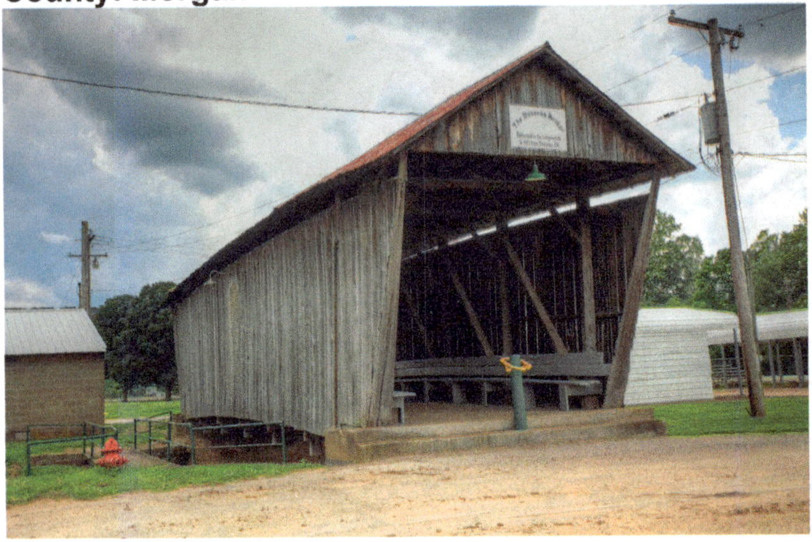

GPS Position: N 39° 38.379' W 81° 50.413'
Directions: Found in the County Fairgrounds at the south end of McConnelsville beside OH-376/Riverside Dr.
Crosses: Ditch
Carries: None
Builder: Not known
Year Built: 1881 (M1953)
Truss Type: Multiple Kingpost
Dimensions: 1 Span, 58 feet

Notes: The bridge was relocated from Rosseau, Ohio spanning Wolf Creek to its present site in 1953. There are park benches in the interior to provide a shady place to sit for Fair visitors.

World Index Number: OH/35-58-32
National Register of Historic Places: Not listed

Island Run (Helmick Mill) Covered Bridge

Township: Deerfield
County: Morgan

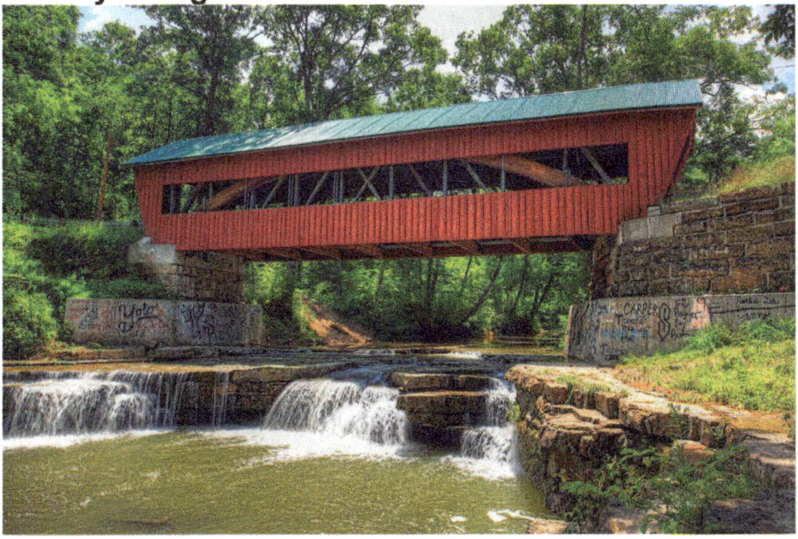

GPS Position: N 39° 43.182' W 81° 56.502'
Directions: From McConnelsville, go north on Township Rd. 954/Curry Lane for 5.4 mi and turn left on CR-74/ Lemon Hill Rd and continue for 3.6 mi. Turn right on Township Road 269/Helmick Road and you will find the bridge in 0.2 miles.
Crosses: Island Run
Carries: Helmick Road
Builder: Not known
Year Built: 1867 (R1940s) (R1991) (R2008)
Truss Type: Multiple Kingpost
Dimensions: 1 Span, 72 feet

Notes: A center pier was added in the 1940s and later, metal siding. The 2008 renovation has the bridge in beautiful shape.

World Index Number: OH/35-58-35
National Register of Historic Places: February 5, 1999

Adams (San Toy) Covered Bridge
Township: Union
County: Morgan

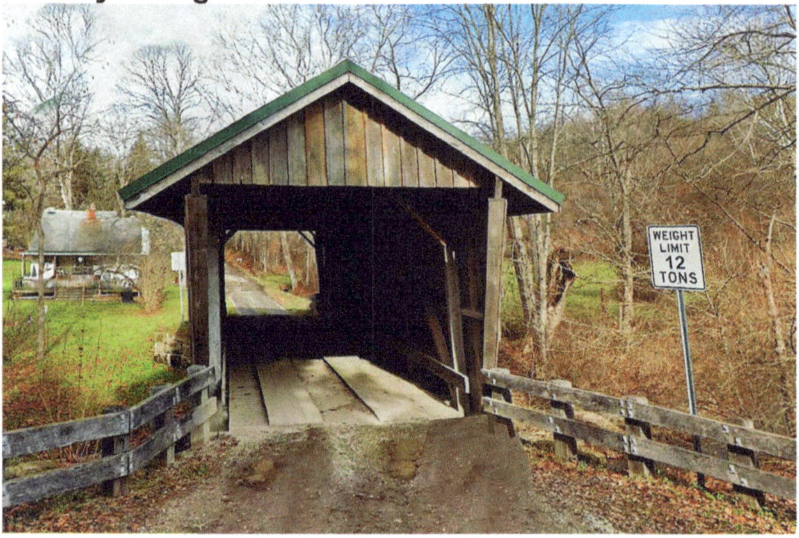

GPS Position: N 39° 37.487' W 82° 01.925'
Directions: From Chapel Hill, head east on Town Hwy 455 for 1.2 miles and turn left onto Morganville Rd and the bridge
Crosses: Santoy Creek
Carries: CR-16/Morganville Rd
Builder: Not known
Year Built: 1875 (R1994) (R2013)
Truss Type: Multiple Kingpost
Dimensions: 1 Span, 57 feet

Notes: This bridge had been closed to traffic for over 10 years and had been in poor condition. In May of 2013 it was renovated and reopened to traffic .

World Index Number: OH/35-58-38
National Register of Historic Places: February 5, 1999

Milton Dye (Saw Mill) Covered Bridge
Township: Manchester
County: Morgan

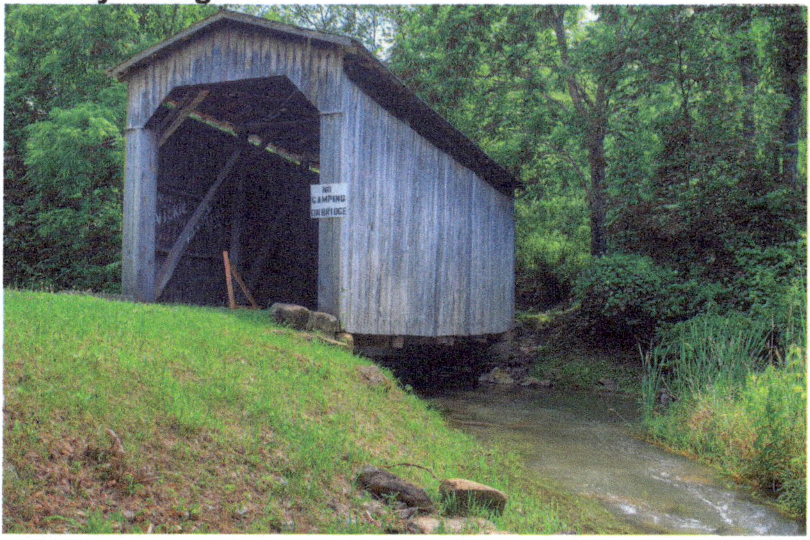

GPS Position: N 39° 44.146' W 81° 41.394'
Directions: From Bristol, go north on OH-78 for 0.4 mi and turn left on OH-83 and then in 0.2 miles a slight left to stay on OH-83. After 3.9 miles turn right onto Sawmill Rd. Continue 0.6 mi onto the Campground Loop where you find the bridge.
Crosses: Brannon's Fork
Carries: By Campground Loop Road
Builder: Milton Dye
Year Built: Ca.1925 (M1965)
Truss Type: Multiple Kingpost
Dimensions: 1 Span, 41 feet

Notes: The bridge originally was located in Noble County but was moved to its present location in 1965. It is closed to traffic

World Index Number: OH/35-58-41
National Register of Historic Places: Not listed

Johnson's Mill (Salt Creek) Covered Bridge

Township: Perry
County: Muskingum

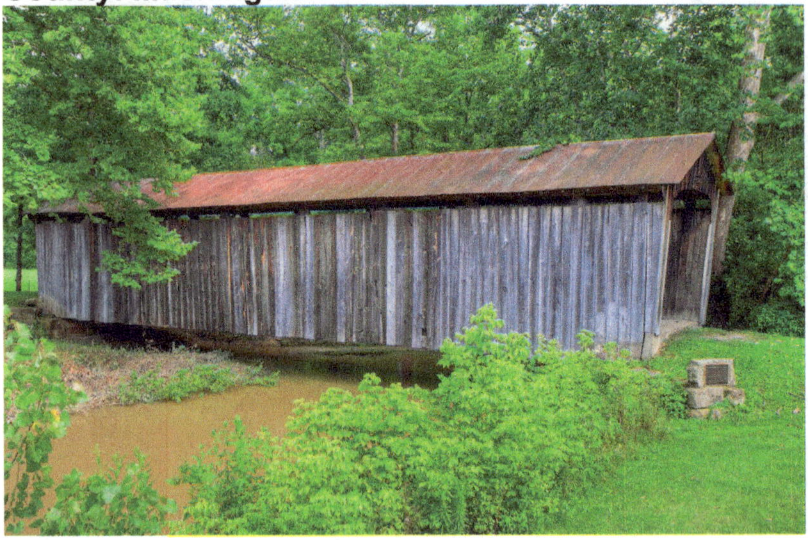

GPS Position: N 39° 59.962' W 81° 50.369'
Directions: From I-70 east of Zanesville take Exit 164 for US-40/US-22 and in 0.2 miles turn right and continue on US-22/US-40 W. In 0.8 miles turn right on Arch Hill Rd and you will find the bridge in 2.2 miles
Crosses: Salt Creek
Carries: Arch Hill Road
Builder: Thomas Fisher
Year Built: 1876 (R1998)
Truss Type: Smith Single
Dimensions: 1 Span, 87 feet

Notes: The bridge is closed to traffic. Formerly thought to have a Warren Truss butt now deemed a Smith Single.

World Index Number: OH/35-60-31
National Register of Historic Places: September 10, 1974

Manchester (Wiley) Covered Bridge

Township: Sharon
County: Noble

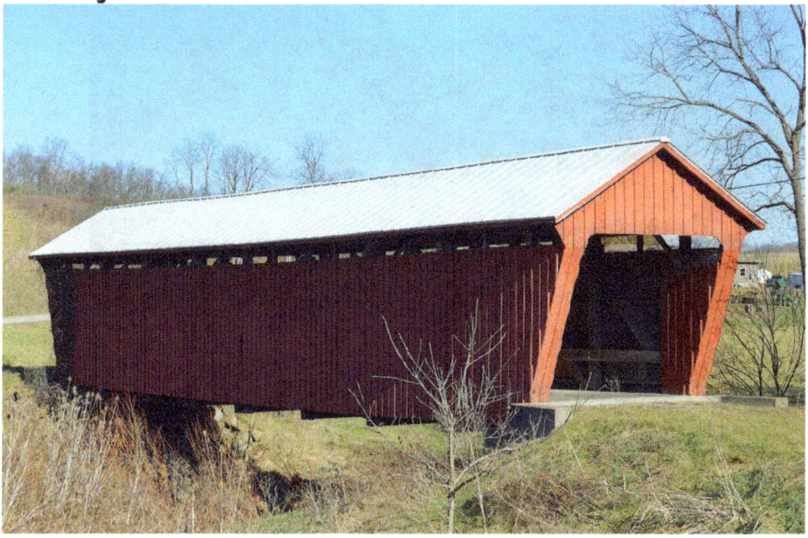

GPS Position: N 39° 42.984' W 81° 37.006'
Directions: Near Caldwell go west on OH-78 from I-77 for 7.0 mi, take a slight left on Hwy 27 and continue on Township Rd 38/Don Wiley Road where you will find the site in 2.7 miles.
Crosses: Olive Green Creek
Carries: Township Rd 38/Don Wiley Road (Bypassed)
Builder: Not known
Year Built: 1915 (R2005)
Truss Type: Multiple Kingpost
Dimensions: 1 Span, 49 feet
Notes: This bridge was built to replace one that was destroyed in the flood of 1913. It was the last covered bridge in Noble County to carry traffic but has been closed and bypassed.

World Index Number: OH/35-61-33
National Register of Historic Places: Not listed

Parrish Covered Bridge
Township: Sharon
County: Noble

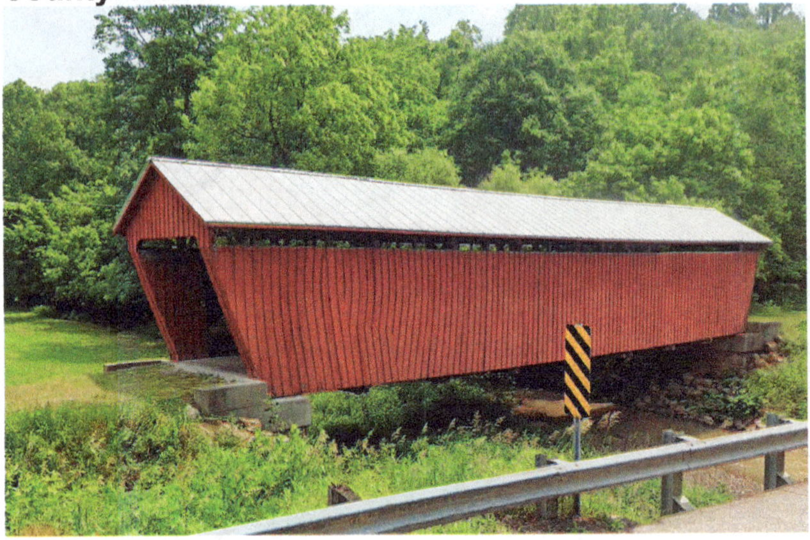

GPS Position: N 39° 42.366' W 81° 34.964'
Directions: From Sharon, go south on Rich Valley Rd where you will find the bridge on a bypassed section in 2.4 miles.
Crosses: Sharon Fork
Carries: Rich Valley Road
Builder: Not known
Year Built: 1914 (R2005)
Truss Type: Multiple Kingpost
Dimensions: 1 plus Span, 81 feet

Notes: This structure was built to replace one destroyed in a 1913 flood. It was closed and bypassed in 1980. A small park has been made around it and it looks good after the 2005 renovation.

World Index Number: OH/35-61-34
National Register of Historic Places: Not listed

Cladwell Fairgrounds Covered Bridge

Township: Olive
County: Noble

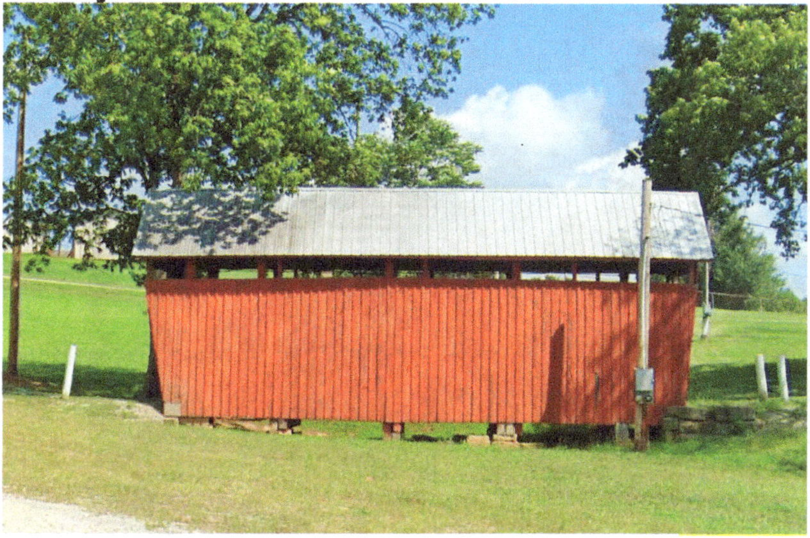

GPS Position: 39°44'32.2"N 81°31'51.5"W
Directions: 16821 Bronze Heights Ln, Caldwell
Crosses: Swale
Carries: Pedestrian walkway

Builder: Not known
Year Built: 1970 (M1970)
Truss Type: Multiple King
Dimensions: 1 Span, 44 feet

Notes: Moved to this location in 1970. Formerly called Park Hill Covered Bridge

World Index Number: OH/35-61-40
National Register of Historic Places: Not listed

Parks (South) Covered Bridge
Township: Hopewell
County: Perry

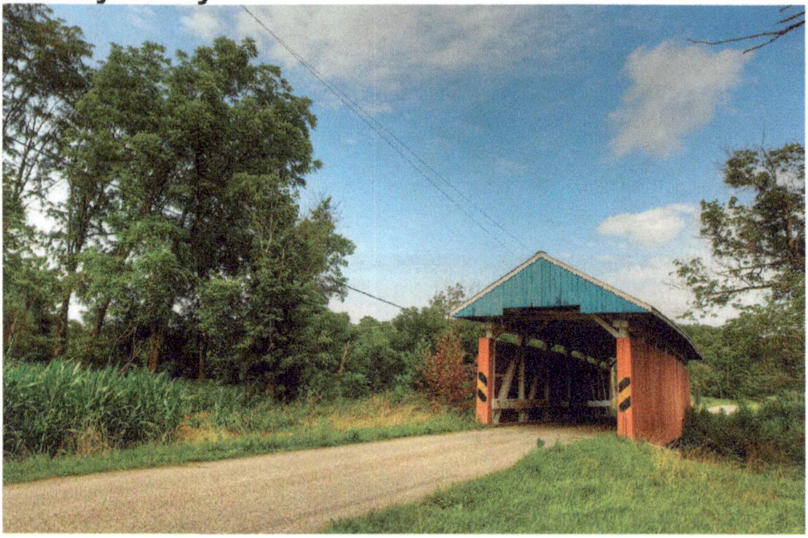

GPS Position: N 39° 51.141' W 82° 16.792'
Directions: From Somerset go north on OH-668 for 2.4 mi and turn left on Gower Rd where the bridge is 1.0 mi.
Crosses: Painter Creek
Carries: Gower Rd
Builder: William A. Dean
Year Built: 1883
Truss Type: Multiple Kingpost
Dimensions: 1 Span, 58 feet

Notes: Looks very similar to nearby Hopewell Church Covered Bridge. A steel stringer supports the structure, and the original kingpost trusses are no longer functional.

World Index Number: OH/35-64-02
National Register of Historic Places: September 10, 1974

Hopewell Church Covered Bridge
Township: Hopewell
County: Perry

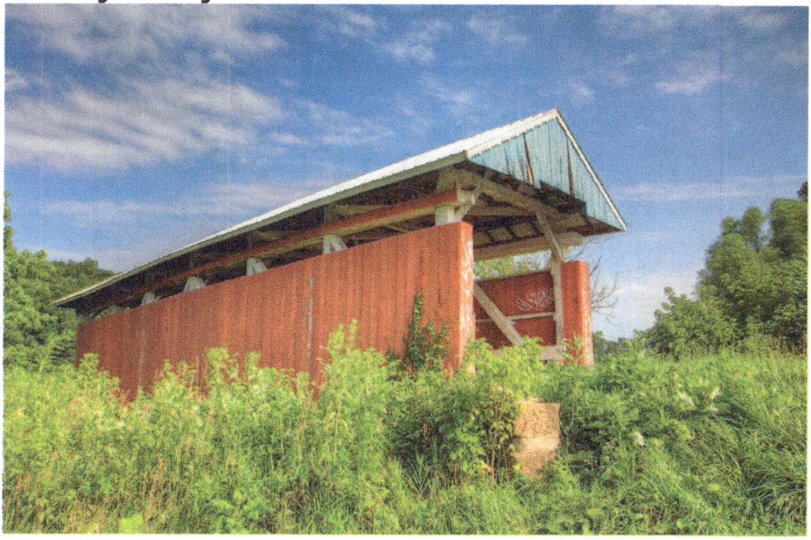

GPS Position: N 39° 51.823' W 82° 16.956'
Directions: From Somerset go north on OH-668 for 2.6 mi and turn left on Township Rd 92 and after 0.7 mi, turn left on CR-51/Copperriders Rd. where you see the bridge in 1.1 mi.
Crosses: Jonathon Creek
Carries: Copperriders Rd.
Builder: Not known
Year Built: 1874 (R2004)
Truss Type: Multiple Kingpost
Dimensions: 1 Span, 54 feet

Notes: The 2004 repair was necessary after vandals tried to burn it. They were caught and imprisoned and neighbors put out the fire. It suffers from wear and graffiti.

World Index Number: OH/35-64-03
National Register of Historic Places: Not listed

Jack's Hollow (Dollings) Covered Bridge
Township: Madison
County: Perry

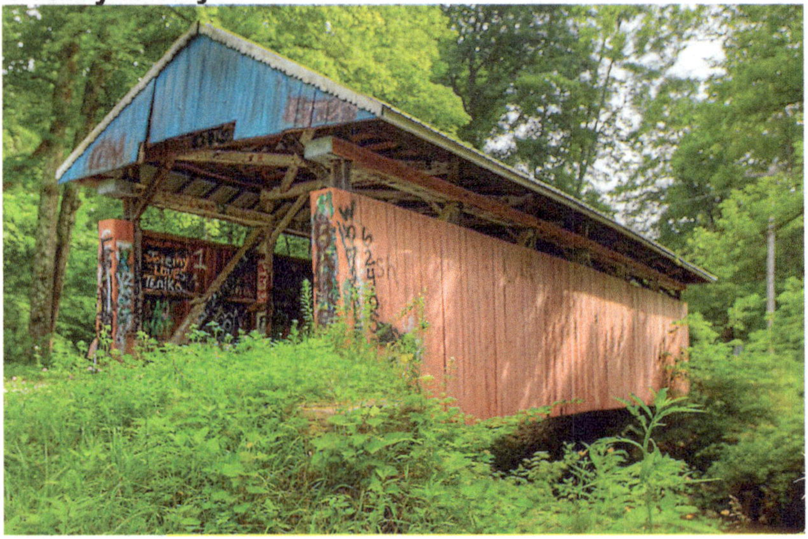

GPS Position: N 39° 54.443' W 82° 10.101'
Directions: From Mt. Perry go north on CR-34/Coppermill Rd. for 1.7 mi and turn right on CR-67/Krofft Rd. Drive 1.9 miles and turn left on Township Hwy 108. The bridge is 0.8 mi.
Crosses: Kent's Run
Carries: Township Hwy 108
Builder: William A. Dean
Year Built: 1879 (R1990s)
Truss Type: Multiple Kingpost
Dimensions: 1 Span, 60 feet

Notes: The bridge is still in use and after vandals and wear, is looking a bit rough. The 1990s repair was needed after vandals set fire to it.

World Index Number: OH/35-64-05
National Register of Historic Places: Not listed

Bowman Hill (Redington) Covered Bridge
Township: Reading
County: Perry

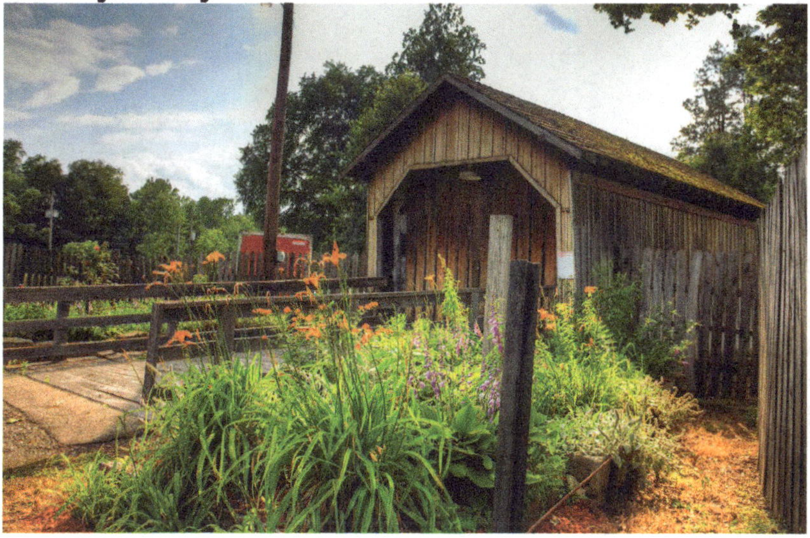

GPS Position: N 39° 43.252' W 82° 13.459'
Directions: At the west end of the town of New Lexington, found off OH-37/Broadway St W at the entrance to the Perry County Fairgrounds
Crosses: None
Carries: Park entrance road, pedestrian only
Builder: Not known
Year Built: 1859 (R1879) (R1884) (R1987) (M1987)
Truss Type: Multiple Kingpost
Dimensions: 1 plus Span, 81 feet

Notes: The bridge was originally located in Reading Township and was moved to its present location in 1987. Its portal is gated and only open when the fairground is in use.

World Index Number: OH/35-64-06
National Register of Historic Places: February 8, 1978

Mary Ruffner (Moore) Covered Bridge

Township: Thorn
County: Perry

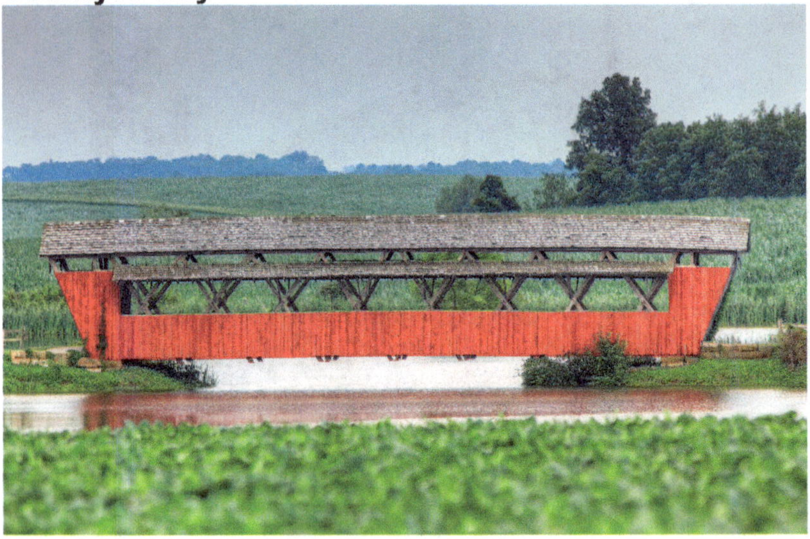

GPS Position: N 39° 51.067' W 82° 23.702'
Directions: From the town of Thornville go south on CR-30/OH-13 for 2.9 miles and the bridge is on the east side of the road near a pond on a private farm
Crosses: Pond
Carries: Private road
Builder: Not known
Year Built: 1875 (M1986)
Truss Type: Smith Double
Dimensions: 1 Span, 78 feet

Notes: The bridge originally crossed Little Rush Creek in Fairfield County. It was moved in 1986 and looks to have been kept in excellent shape.

World Index Number: OH/35-64-84
National Register of Historic Places: Not listed

Valentine (Bill Green) Covered Bridge
Township: Scioto
County: Pickaway

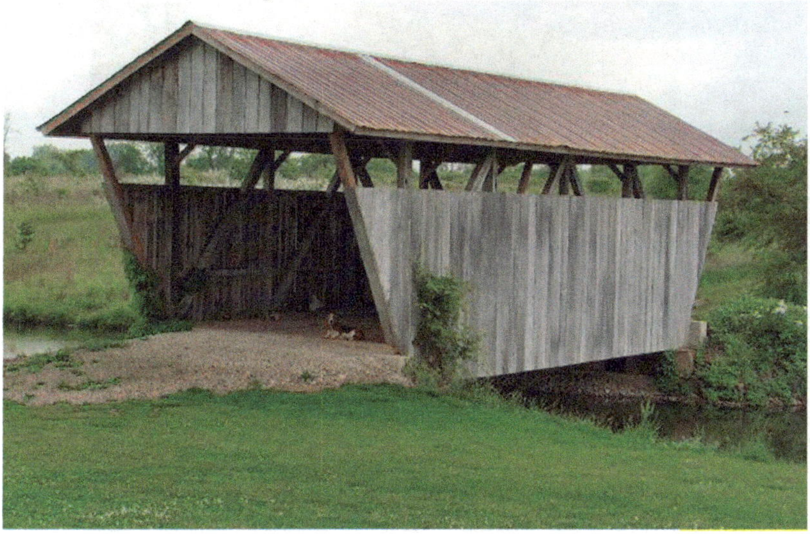

GPS Position: N 39° 46.476' W 83° 07.651'
Directions: Found on a private farm west of the town of Commercial Point
Crosses: Pond
Carries: Private road
Builder: Not known
Year Built: 2007
Truss Type: Multiple Kingpost
Dimensions: 1 Span, 36 feet

Notes: A bridge known as the Green Covered Bridge was moved to this site in 1978. It was blown down in 2004 but the present bridge built in its place in 2007. Private, permission required

World Index Number: OH/35-65-15#2
National Register of Historic Places: Not listed

Blackburn (Brannon, Wesner) Covered Bridge
Township: Madison
County: Pickaway

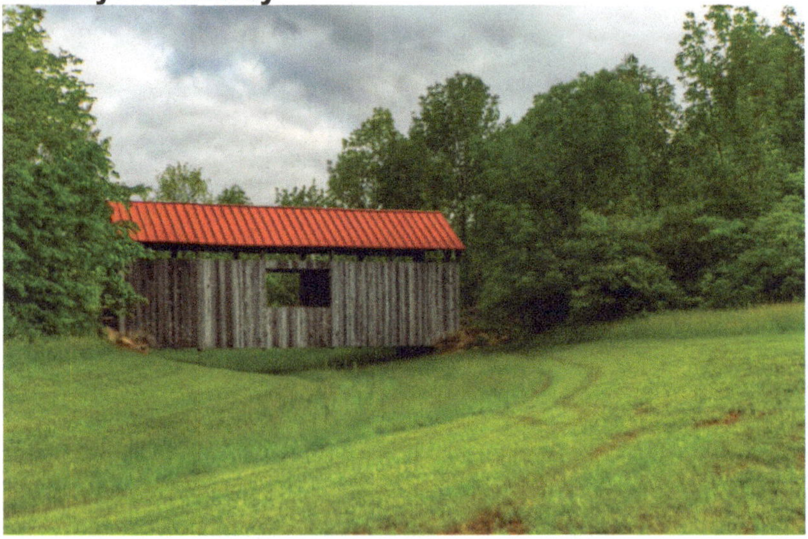

GPS Position: N 39° 45.715' W 82° 50.874'
Directions: Found by an interior park road in Slate Run Metro Park south of the town of Lithopolis
Crosses: Slate Run
Carries: N/A
Builder: Not known
Year Built: 1885 (M1967) (M1999)
Truss Type: Multiple Kingpost
Dimensions: 1 Span, 60 feet

Notes: The bridge initially spanned Brannon's Fork in Muskinhum County (designated 35-60-28) and was moved in 1967 because the area was strip mined. It was moved to its present location in 1999.

World Index Number: OH/35-65-16
National Register of Historic Places: Not listed

Harshman Covered Bridge
Township: Dixon
County: Preble

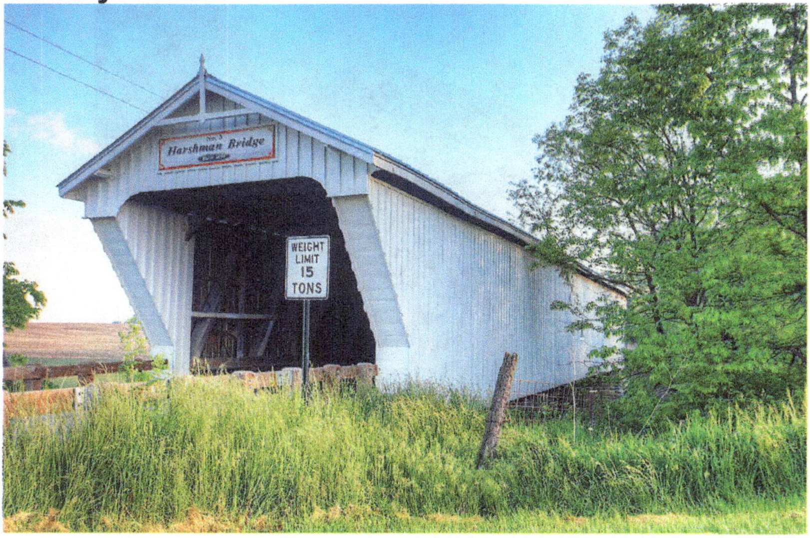

GPS Position: N 39° 42.129' W 84° 46.175'
Directions: From Eaton, go west on OH-122 for 4.3 mi and turn left on Wyatt Rd. After 3.0 mi, turn right on California School Rd. and after 1.6 mi left on Concord Fairhaven Rd where you will find the bridge in 0.2 miles.
Crosses: Four Mile Creek
Carries: Concord Fairhaven Rd
Builder: Everett S. Sherman
Year Built: 1894 (R2008)
Truss Type: Childs
Dimensions: 1 Span, 104 feet
Notes: The 2008 rehabilitation gave the building a new floor, roof and sides. A number of chords were added and the load capacity was brought from 6 to 15 tons. It looks great.

World Index Number: OH/35-68-03
National Register of Historic Places: September 29, 1976

Dixon Branch Covered Bridge

Township: Harrison
County: Preble

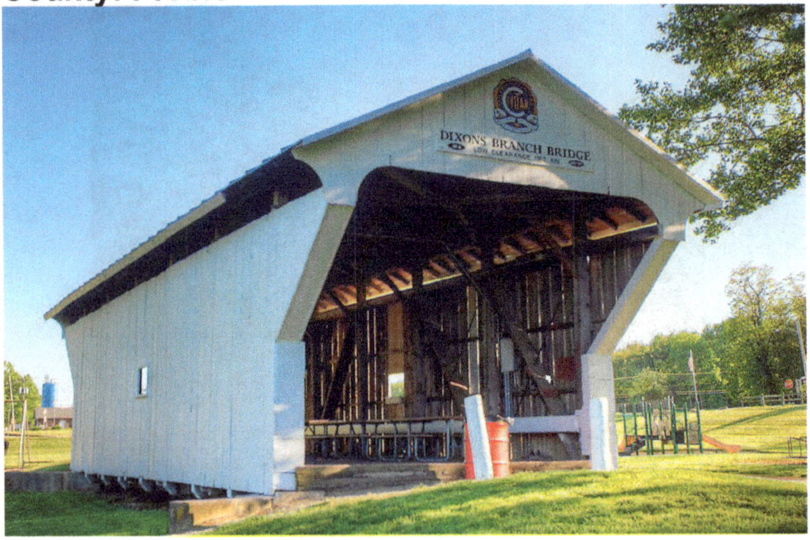

GPS Position: N 39° 50.743' W 84° 32.135'
Directions: In the town of Lewisburg go east on Dayton St/Salem Rd off OH-503 and turn left into Lewisburg Community Park where you will see the bridge.
Crosses: None
Carries: N/A
Builder: Everett S. Sherman
Year Built: 1887 (R1964) (M1964)
Truss Type: Childs
Dimensions: 1 Span, 50 feet
Notes: The bridge was originally located on the Concord Road crossing Dixon's Branch. In 1963 it was damaged in a windstorm. It was repaired and moved to its present site in 1964.

World Index Number: OH/35-68-04
National Register of Historic Places: Not listed

Roberts Covered Bridge
Township: Eaton
County: Preble

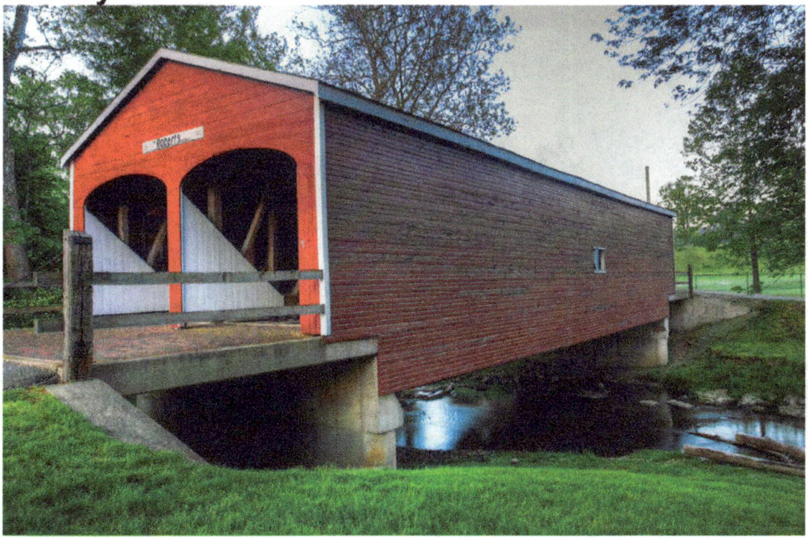

GPS Position: N 39° 44.428' W 84° 38.291'
Directions: In the Town of Eaton, found by Beach St. between St. Clair St. and Israel St.
Crosses: Seven Mile Creek
Carries: N/A
Builder: Orlistus Roberts and James Campbell
Year Built: 1829 (R1991) (M1991)
Truss Type: Burr Arch
Dimensions: 1 Span, 79 feet
Notes: One of only a few Double Barreled (2 separated lanes) and the oldest surviving, it was hit by arson in 1986 but was restored and moved to its present location in 1991. It originally stood on Old Camden Road. This is Ohio's oldest surviving covered bridge.

World Index Number: OH/35-68-05
National Register of Historic Places: September 3, 1971

Brubaker Covered Bridge

Township: Gratis
County: Preble

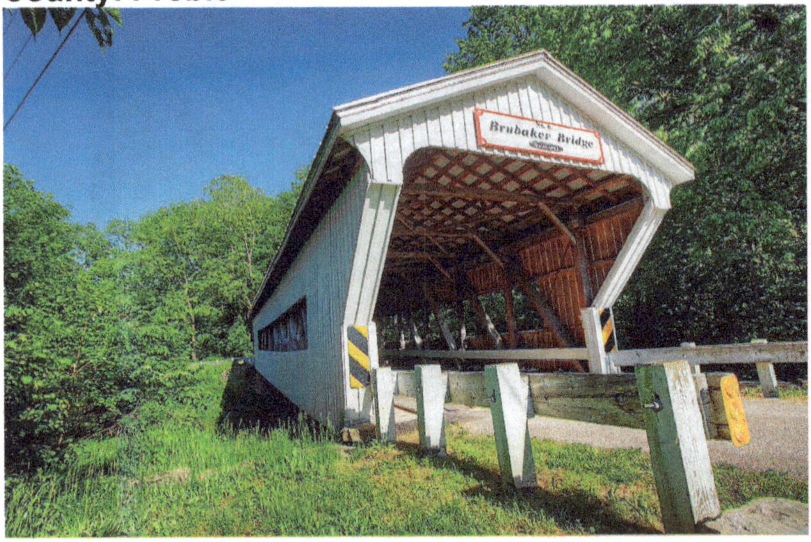

GPS Position: N 39° 39.095' W 84° 32.610'
Directions: From Gratis go west on OH-122 and after 0.3 mi, turn left on Brubaker Rd. And the bridge is 0.6 miles
Crosses: Sam's Run
Carries: Brubaker Road
Builder: Everett S. Sherman
Year Built: 1887 (2006)
Truss Type: Childs
Dimensions: 1 Span, 85 feet

Notes: The open sides are a later addition which allow improved sight lines because of a sharp road curve. The 2006 renovation raised the load limit to 12 tons.

World Index Number: OH/35-68-06
National Register of Historic Places: June 11, 1975

Christman Covered Bridge
Township: Washington
County: Preble

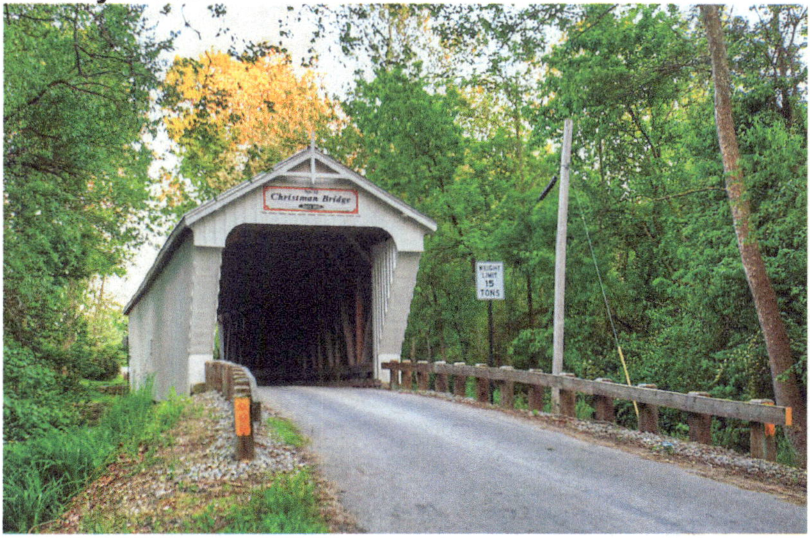

GPS Position: N 39° 46.236' W 84° 39.343'
Directions: From Eaton, go northwest on US-35 for 1.5 mi and turn right on Washington Jackson St. After 1.2 mi go left on Easy Gettysburg Rd. and in 0.4 miles, a slight left on Eaton New Hope Road. You will find the bridge after 0.2 miles.
Crosses: Seven Mile Creek
Carries: Eaton New Hope Road
Builder: Everett S. Sherman
Year Built: 1895 (R1920) (R1960) (R1961)
Truss Type: Childs
Dimensions: 1 Span, 92 feet

Notes: The structure still carries vehicle traffic and was built to provide access to a sawmill.

World Index Number: OH/35-68-12
National Register of Historic Places: October 22, 1976

Geeting Covered Bridge
Township: Monroe
County: Preble

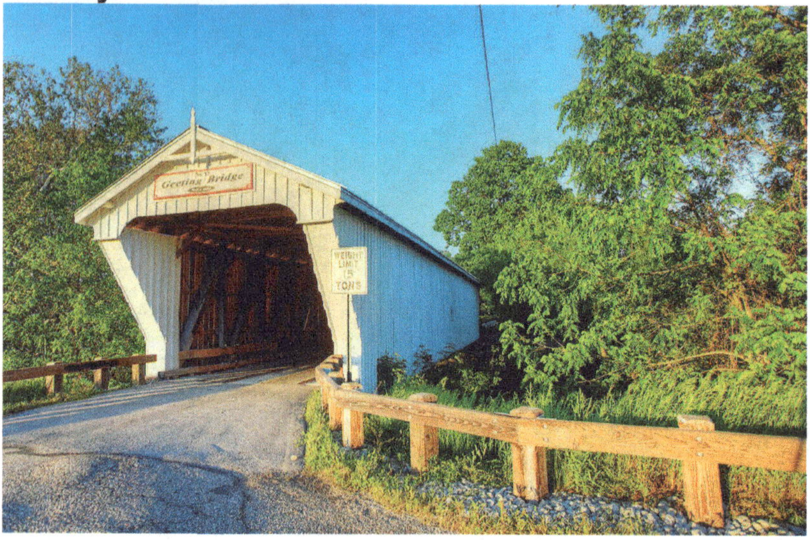

GPS Position: N 39° 50.611' W 84° 35.842'
Directions: From the town of Lewisburg, go west on US-40 for 2.2 miles and turn right on Yohe Rd. In 0.6 miles, turn right on Price Rd. where the bridge is a short distance.
Crosses: Price Creek
Carries: Price Rd.
Builder: Everett S. Sherman
Year Built: 1894 (R1973) (R2008)
Truss Type: Childs
Dimensions: 1 Span, 100 feet

Notes: The 2008 repairs has left this bridge in good shape. The Childs truss was revived by the builder after its patent had expired.

World Index Number: OH/35-68-13
National Register of Historic Places: August 19, 1975

Warnke Covered Bridge
Township: Monroe
County: Preble County

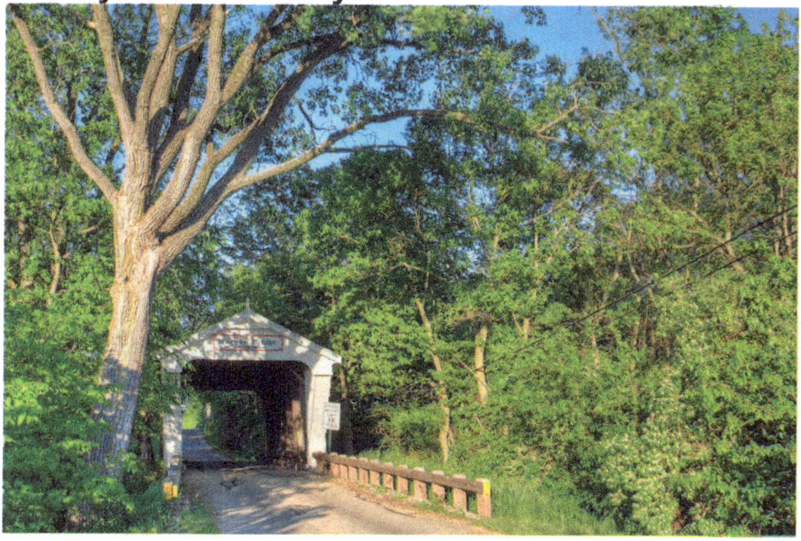

GPS Position: N 39° 52.420' W 84° 30.906'
Directions: From Lewisburg, go east on US-40 for 0.2 miles and turn left on Euphonia Verona Rd. After 1.5 miles, turn right on Swamp Creek Rd. where the bridge is a short distance
Crosses: Swamp Creek
Carries: Swamp Creek Rd.
Builder: E.S. Sherman
Year Built: 1895 (R1913) (R2008)
Truss Type: Childs
Dimensions: 1 Span, 51 feet

Notes: This is said to be the last bridge built by E.S. Sherman, who has 7 bridges surviving in Ohio and 6 in Preble County. It replaced a steel truss bridge that was damaged in a flood

World Index Number: OH/35-68-14
National Register of Historic Places: October 8, 1976

Hueston Woods Covered Bridge
Township: Israel
County: Preble County

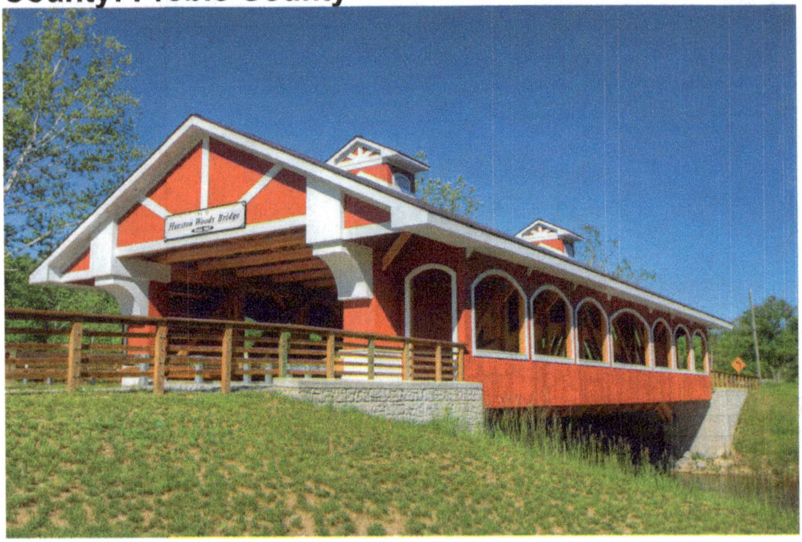

GPS Position: N 39° 35.515' W 84° 46.192'
Directions: From West College Corner go north on Camden College Corner Rd and you will arrive at the bridge after 1.5 mi
Crosses: Four Mile Creek
Carries: Camden College Corner Rd

Builder: Brumbaugh Construction
Year Built: 2012
Truss Type: Double Burr Arch
Dimensions: 1 Span, 108 feet

Notes: This beautiful new bridge carries two lanes of traffic as well as two pedestrian walkways. It also has two copulas which are an interesting design elements.

World Index Number: OH/35-68-51
National Register of Historic Places: Not listed

Buckskin (South Salem) Covered Bridge
Township: Buckskin
County: Ross

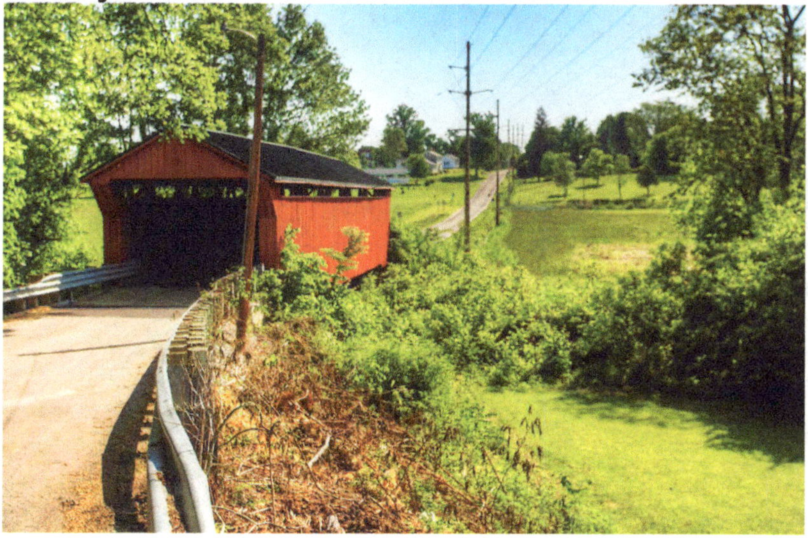

GPS Position: N 39° 20.027' W 83° 18.918'
Directions: Just east of the town of South Salem on Lower Twin Road.
Crosses: Buckskin Creek
Carries: Lower Twin Road
Builder: Smith Bridge Company
Year Built: 1873 (R2006) (R2013)
Truss Type: Smith Double
Dimensions: 1 Span, 99 feet

Notes: A new roof, floor and sides were added in the 2006 renovation. It is in active use. It is the last surviving covered bridge in the county. In the late 1990s there were plans for its demolition but community support saved it.

World Index Number: OH/35-71-02
National Register of Historic Places: March 4, 1975

Mull Road Covered Bridge

Township: Ballville
County: Sandusky

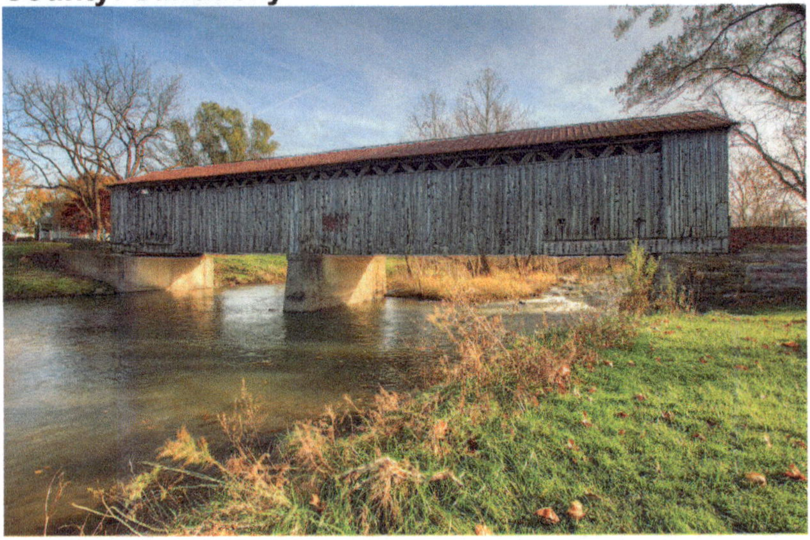

GPS Position: N 41° 15.658' W 83° 11.066'
Directions: From Fremont go south on Cr-132/River Road for 2.6 miles and turn right on Olc CR-35. In 0.2 miles turn left on OH-53 and drive 2.0 miles where you turn right on CR-9. In 0.9 miles you will find the bridge on a bypassed section..
Crosses: Wolf Creek, East Branch
Carries: County Road 9 (Bypassed section)
Builder: Henry Mull Family
Year Built: 1842
Truss Type: Town
Dimensions: 1 plus Span, 99 feet
Notes: The bridge was built by the Mull family to provide access to their mill. It was closed to traffic and bypassed when the new bridge was completed in 1962.

World Index Number: OH/35-72-01
National Register of Historic Places: October 15, 1974

Otway Covered Bridge

Township: Brush Creek
County: Scioto

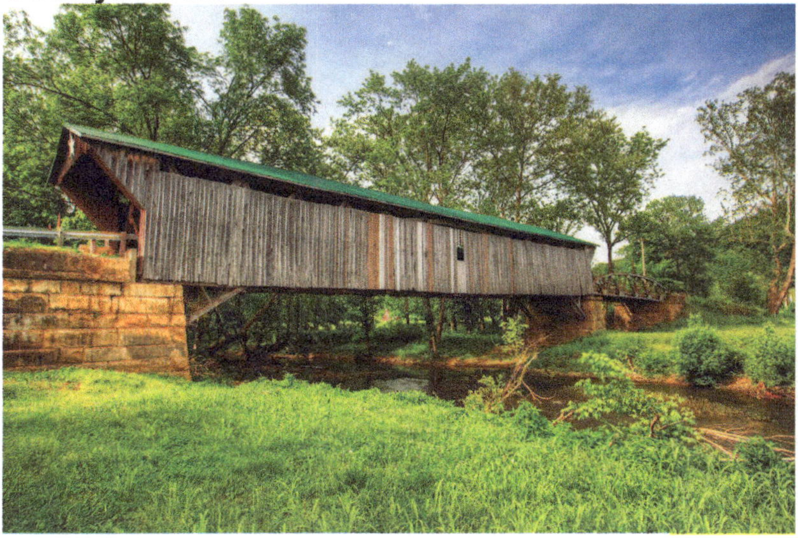

GPS Position: N 38° 51.742' W 83° 11.443'
Directions: From Otway go southwest on OH-348 for 0.5 miles and you will find the bridge on a bypassed section.
Crosses: Scioto Brush Creek
Carries: OH-348 (Bypassed Section)

Builder: Smith Bridge Co.
Year Built: 1874 (R1896) (1920s) (R1974) (R2014)
Truss Type: Smith Double and arch
Dimensions: 1 Span, 127 feet

Notes: The Burr arch was added in 1896. The bridge was closed and bypassed in 1963. It was completely restored by the village of Otway in 1974.

World Index Number: OH/35-73-15
National Register of Historic Places: May 3, 1974

Everett Road Covered Bridge

Township: Boston
County: Summit

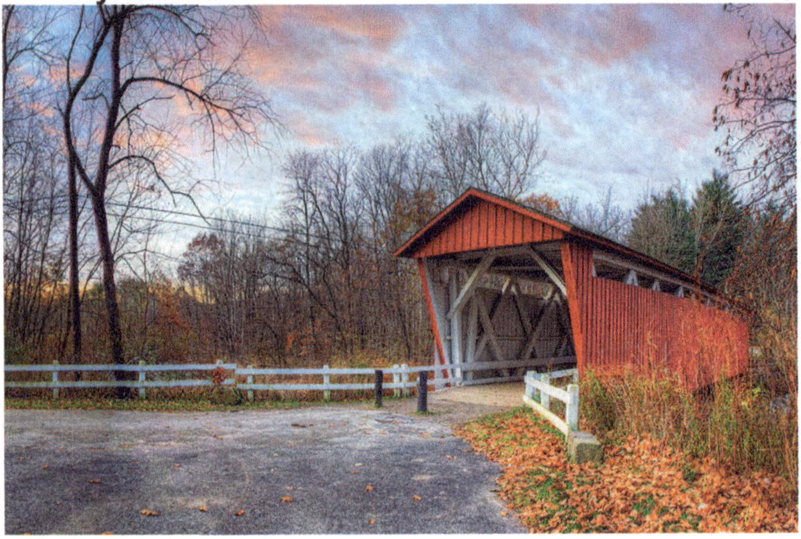

GPS Position: N 41° 12.223' W 81° 35.031'
Directions: From Everett, take Everett Rd. for 1.5 mi west and find the bridge at the intersection with Oak Hill Rd.
Crosses: Furnace Run
Carries: Everett Road (Bypassed)
Builder: Not Known
Year Built: 1986 (Original Ca. 1877)
Truss Type: Smith Double
Dimensions: 1 Span, 100 feet

Notes: The bridge is said to have been built after a man drowned while trying to cross the stream in 1877. The bridge was swept away and destroyed by a flood in 1975. It was re-built in a historically accurate matter in 1986.

World Index Number: OH/35-77-01#2
National Register of Historic Places: Not listed

Newton Falls (Stedman) Covered Bridge
Township: Newton
County: Trumbull

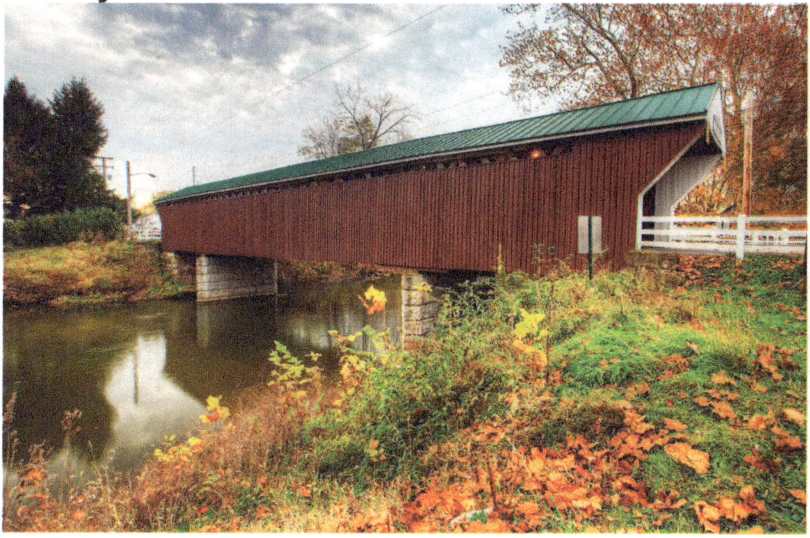

GPS Position: N 41° 11.278' W 80° 58.309'
Directions: In the town of Newton Falls it is found on Arlington Rd. near the junction of River St.
Crosses: Mahoning River, East Branch
Carries: Arlington Rd
Builder: Not known
Year Built: 1831 (R1921) (R1962) (R1985) (R2007) (R2009)
Truss Type: Town
Dimensions: 1 plus Span, 117 feet

Notes: This is Ohio's second oldest covered bridge and the oldest still in use by vehicles. A pedestrian walkway was added in 1921. The repairs in 1985 were needed after tornado damage and those in 2009 after damage from a truck.

World Index Number: OH/35-78-01
National Register of Historic Places: October 16, 1974

Upper Darby (Pottersburg) Covered Bridge

Township: Allen
County: Union

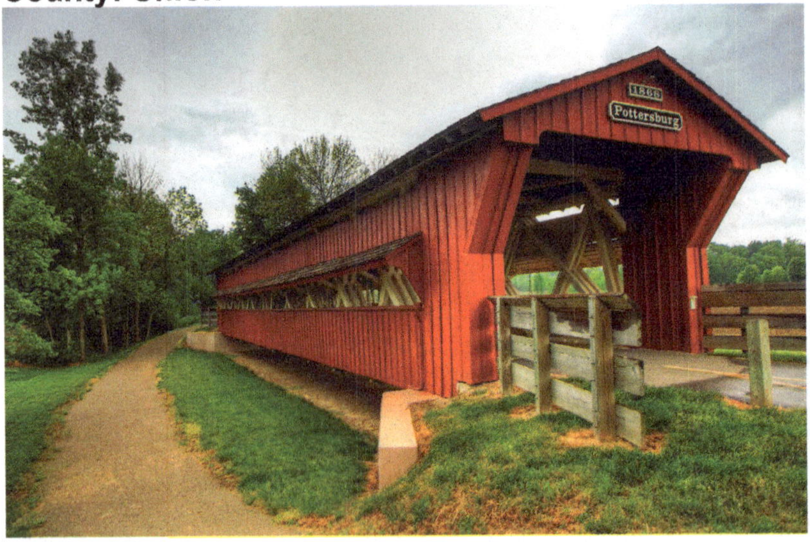

GPS Position: N 40° 14.450' W 83° 31.709'
Directions: From the town of North Lewisburg, go east on OH-245/Maple St. for 1.5 miles and turn left on Inskeep-Cratty Rd where you will see the bridge in 0.7 miles
Crosses: None
Carries: Inskeep-Cratty Rd (Bypassed)
Builder: Rueben Partridge
Year Built: 1868 (M2006)
Truss Type: Partridge
Dimensions: 1 Span, 94 feet
Notes: This bridge was originally located on the North Lewisburg Road were it spanned the Big Darby Creek. It was moved to the present site in 1966 and is currently on a hiking trail.

World Index Number: OH/35-80-01
National Register of Historic Places: Not listed

Spain Creek Covered Bridge
Township: Allen
County: Union

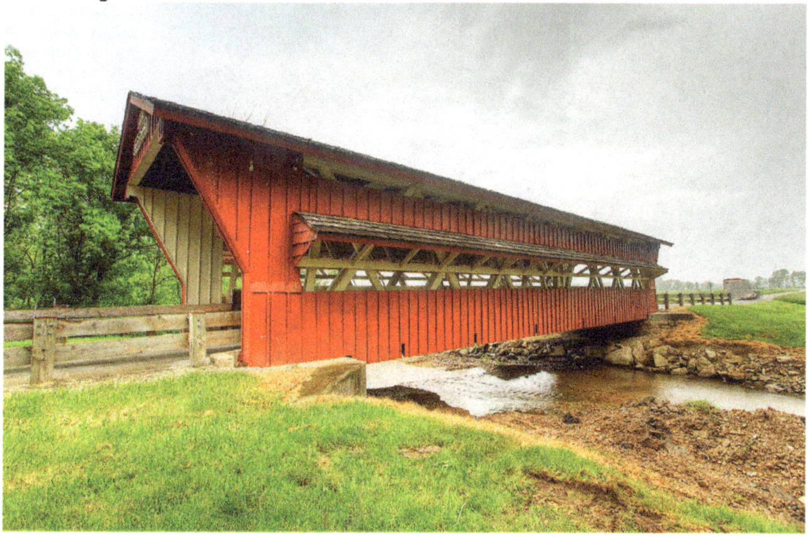

GPS Position: N 40° 13.706' W 83° 31.377'
Directions: From the town of North Lewisburg, go east on OH-245 for 1.4 miles and turn left on Inskeep-Cratty Rd where you will see the bridge.
Crosses: Spain Creek
Carries: Inskeep-Cratty Rd
Builder: Reuben Partridge
Year Built: CA. 1870 (R1988)
Truss Type: Partridge
Dimensions: 1 Span, 75 feet

Notes: The builder, Reuben Partridge built a number of Union county bridges of which 4 survive. The window awnings are a later addition.

World Index Number: OH/35-80-02
National Register of Historic Places: Not listed

Culbertson (Winget Road, Treacle Creek) CB
Township: Union
County: Union

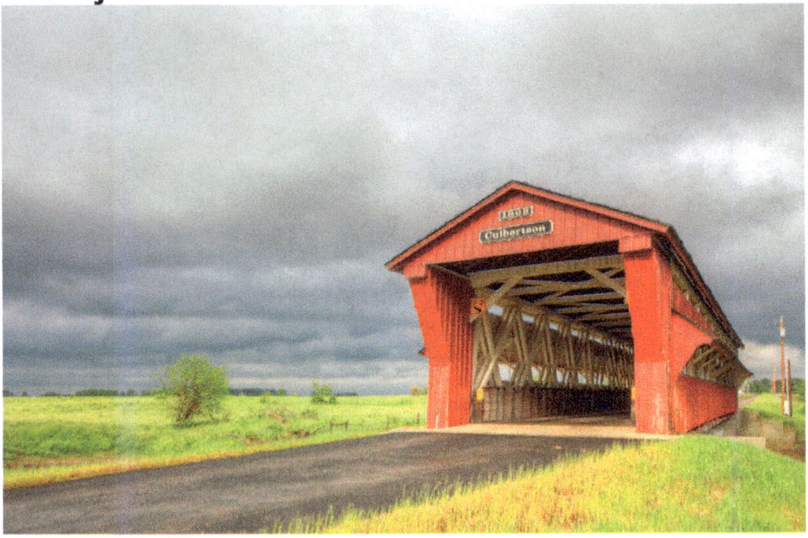

GPS Position: N 40° 08.292' W 83° 27.170'
Directions: From Milford Center, go southwest on OH-4 for 2.3 miles and turn left on Homer Rd. After 0.7 miles turn left on Winget Rd. where you will see the bridge in 0.3 miles.
Crosses: Treacle Creek
Carries: Winget Rd.
Builder: Reuben Partridge
Year Built: 1868 (M1921) (R1988)
Truss Type: Partridge
Dimensions: 1 Span, 100 feet
Notes: Reuben Partridge built this bridge, originally located on State Route 4 south of Milford center. It was rebuilt in its present location in 1921. It is a gorgeous structure in a natural setting.

World Index Number: OH/35-80-03
National Register of Historic Places: Not listed

Bigelow (Little Darby, Axe Handle Road) CB
Township: Union
County: Union

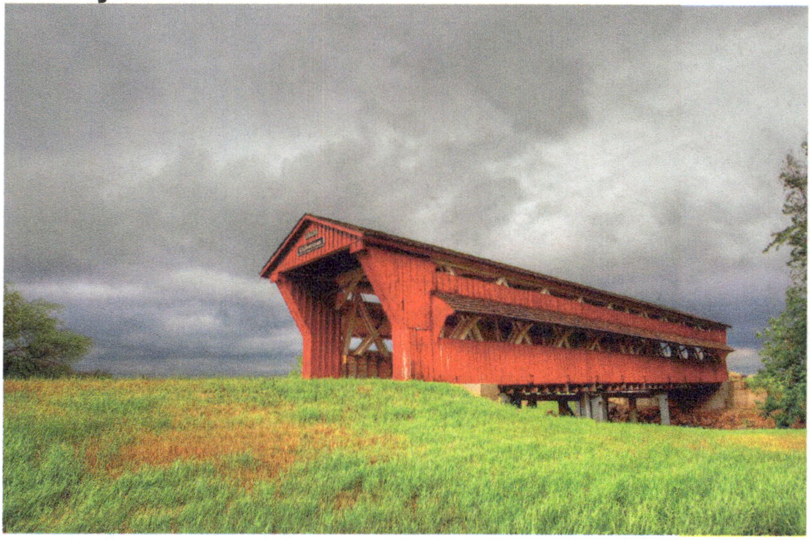

GPS Position: N 40° 06.982' W 83° 25.495'
Directions: From Chuckery go west on OH-161 for 1.8 mi and turn right on Axe Handle Rd. The bridge is a short distance.
Crosses: Little Darby Creek
Carries: Axe Handle Rd.

Builder: Reuben Partridge
Year Built: 1873 (R1991)
Truss Type: Partridge
Dimensions: 1 Span, 114 feet

Notes: One of several Union County bridges built by Reuben Partridge. The window awnings are likely a later addition and the 1991 renovation added steel supports.

World Index Number: OH/35-80-04
National Register of Historic Places: Not listed

New Upper Darby (N Lewisburg Rd, Pottersburg#2)
Township: Allen
County: Union

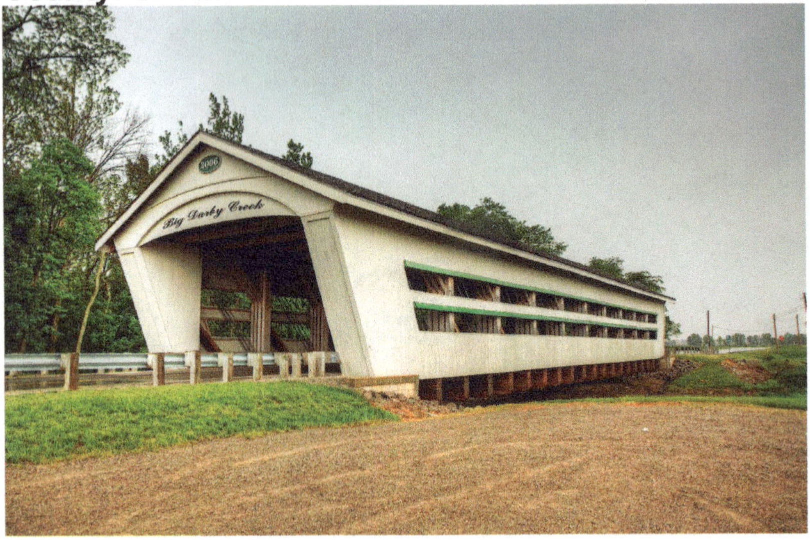

GPS Position: N 40° 14.975' W 83° 31.709'
Directions: From the town of Marysville go south on OH-245 for 6.4 miles and turn right on West Darby Road where you will find the bridge in 1.7 miles
Crosses: Big Darby Creek
Carries: West Darby Road
Builder: Not known
Year Built: 2007
Truss Type: Modified Pratt
Dimensions: 1 Span, 130 feet

Notes: One of Ohio's new Covered bridges designed to bear a heavy traffic load. The double rows of side windows looks great.

World Index Number: OH/35-80-61
National Register of Historic Places: Not listed

Buck Run Road (Big Darby Creek) CB

Township: Allen
County: Union

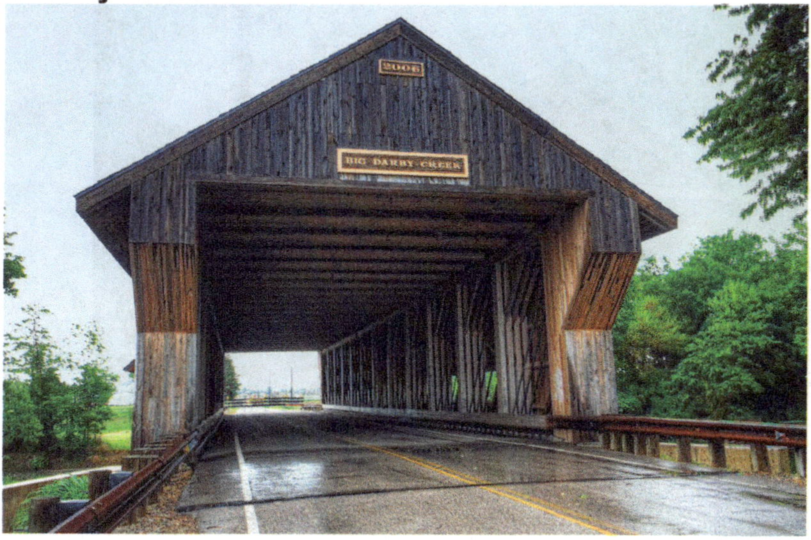

GPS Position: N 40° 12.578' W 83° 29.373'
Directions: From North Lewisburg, go east on OH-245 for 2.8 miles and turn right on Darby Roe Rd. In 1.3 miles turn right on Buck Run Rd where you will find the bridge in 0.3 miles.
Crosses: Big Darby Creek
Carries: Buck Run Rd
Builder: Righter Co
Year Built: 2007
Truss Type: Modified Pratt
Dimensions: 1 Span, 160 feet
Notes: The bridge was found to have the upper chords out of alignment in 2012 and was closed. It has since re-opened. Like many of Ohio's recently built covered bridges, it employs the Pratt Truss system

World Index Number: OH/35-80-62
National Register of Historic Places: Not listed

Mill Creek Covered Bridge

Township: Millcreek
County: Union

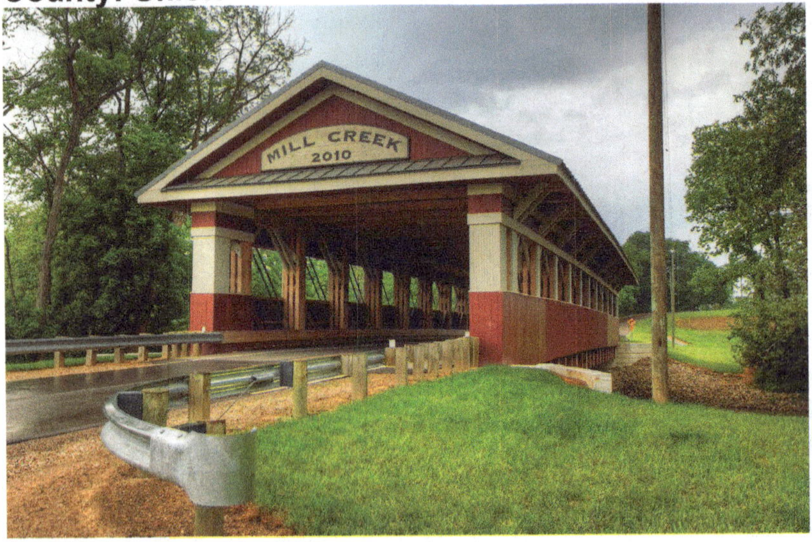

GPS Position: N 40° 14.565' W 83° 13.983'
Directions: From Marysville, go east on US-36 for 6.5 mi and turn right onto CR-157/Dean Rd which continues as Stover Rd and Thompson Rd. In 1.8 miles you will find the bridge.
Crosses: Mill Creek
Carries: Thompson Rd.
Builder: Zachrich Construction
Year Built: 2010
Truss Type: Modified Pratt
Dimensions: 1 Span, 125 feet

Notes: This new bridge replaced an old metal bridge. It was assembled by the side of the road and then moved into place over two days using a system of wheeled dollies and winches.

World Index Number: OH/35-80-63
National Register of Historic Places: Not listed

Mount Olive (Grandstaff) Covered Bridge

Township: Jackson-Richland
County: Vinton

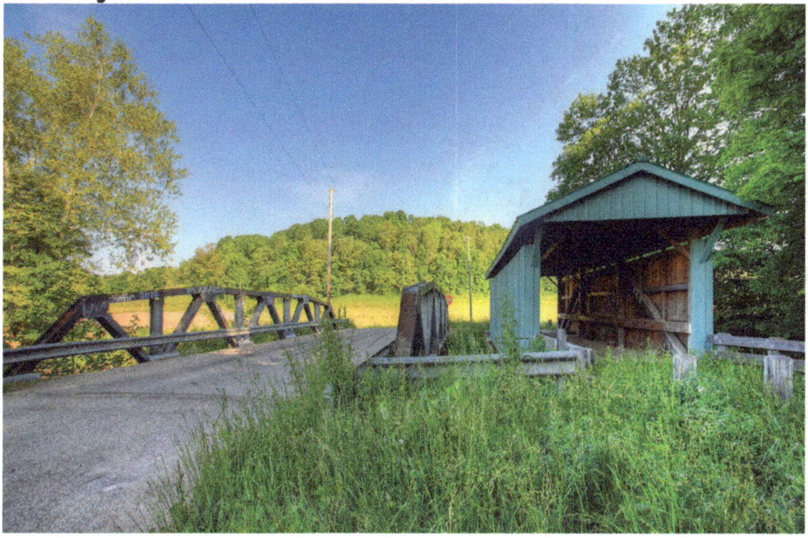

GPS Position: N 39° 17.187' W 82° 35.332'
Directions: From the town of Allensville, go north on Goosecreek Rd. for 1.4 miles and turn right on Mt. Olive Road where you will see the bridge on a bypassed section.
Crosses: Salt Creek, Middle Fork
Carries: Mt. Olive Road (Bypassed Section)

Builder: George Washington Pilcher
Year Built: 1875 (R1930) (R1963) (R1979)
Truss Type: Queen
Dimensions: 1 Span, 48 feet

Notes: When the bridge was built, it lay amid land owned by a family named "Grandstaff"; which led to its alternate name.

World Index Number: OH/35-82-04
National Register of Historic Places: October 8, 1976

Bay (Tinker) Covered Bridge
Township: Elk
County: Vinton

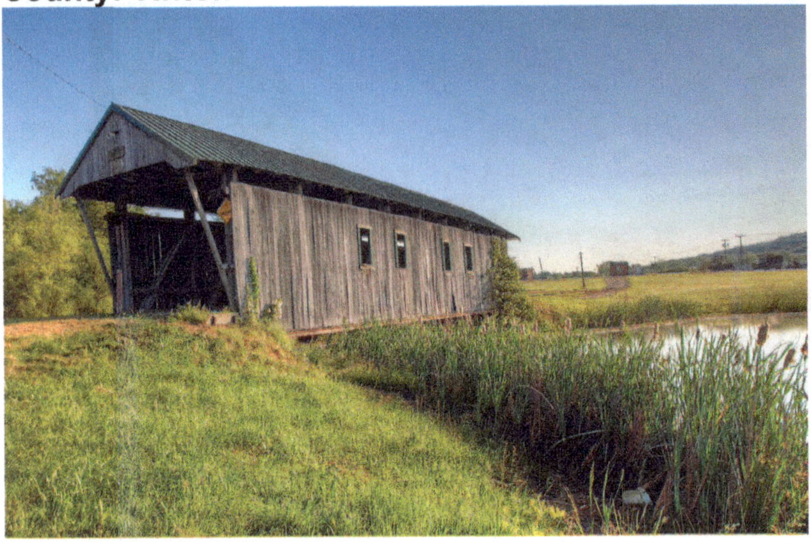

GPS Position: N 39° 16.171' W 82° 28.499'
Directions: from the town of McArthur go north on OH-93 for 1.1 miles where you will see the bridge by the road in the Junior Fairgrounds.
Crosses: Pond
Carries: N/a
Builder: Not known
Year Built: 1876 (M1966) (R2001)
Truss Type: Double Multiple Kingpost
Dimensions: 1 Span, 63 feet

Notes: The bridge was moved 10 miles from its initial site over Little Raccoon Creek when that area was dammed. It has been exhibited by a quiet pond at the fairground site

World Index Number: OH/35-82-05
National Register of Historic Places: Not listed

Eakin Mill (Arbaugh) Covered Bridge
Township: Vinton
County: Vinton

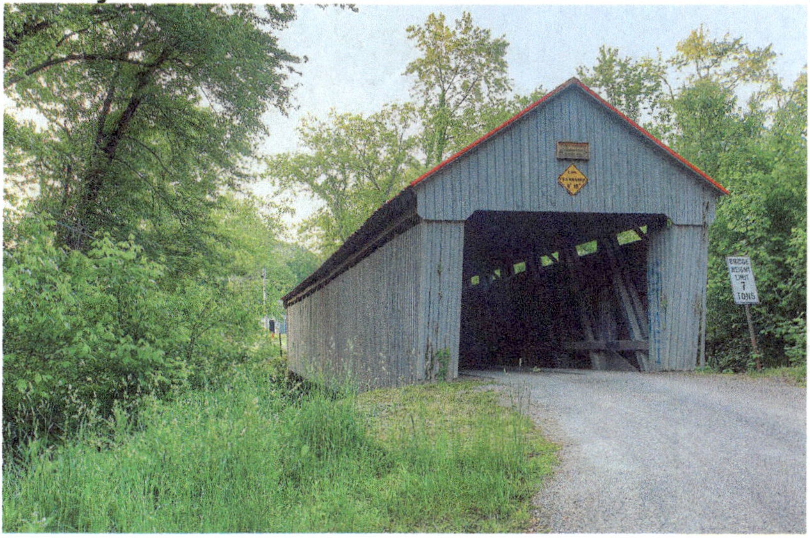

GPS Position: N 39° 10.306' W 82° 20.221'
Directions: From Albany go west on OH-32 and after 8.2 mi turn right on Eakin Mill Road where the site is found in 0.5 mi
Crosses: Raccoon Creek
Carries: Eakin Mill Road

Builder: Not known
Year Built: 2003
Truss Type: Double multiple Kingpost and Arch
Dimensions: 1 Span, 111 feet

Notes: The original 1870 bridge was damaged by a truck and closed. In 2002 it was removed and the present bridge built in its place.

World Index Number: 35-82-07#2
National Register of Historic Places: Not listed

Cox Covered Bridge
Township: Swan
County: Vinton

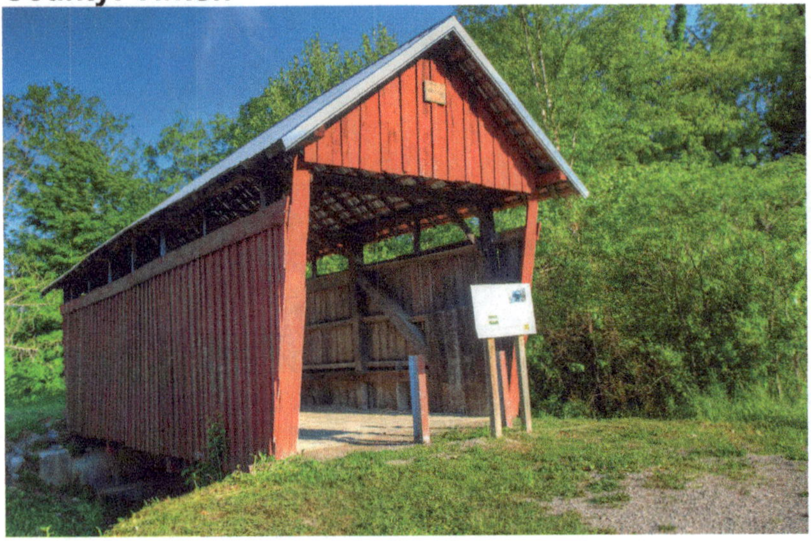

GPS Position: N 39° 21.949' W 82° 27.618'
Directions: From Logan go south on OH-93 for 12.6 miles and turn left on Woodgeard Rd and see the bridge
Crosses: Raccoon Creek, Bushy Fork
Carries: Woodgeard Rd (Bypassed section)
Builder: Diltz and Steele Co.
Year Built: 1884 (M1992) (2004)
Truss Type: Queen
Dimensions: 1 Span, 40 feet

Notes: When it was closed and bypassed in 1992, it was moved 20 feet onto new concrete abutments, in order to make room for a modern replacement. It is now the site of a small picnic area.

World Index Number: OH/35-82-10
National Register of Historic Places: Not listed

Shinn Covered Bridge
Township: Palmer
County: Washington

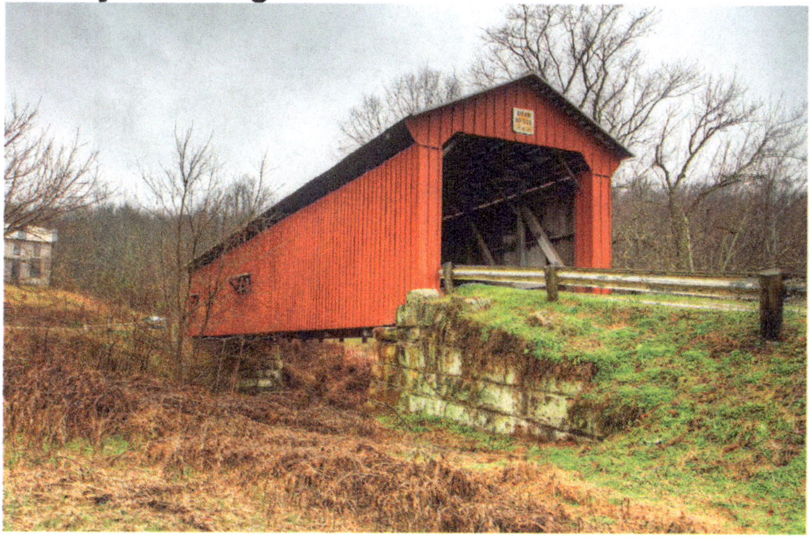

GPS Position: N 39° 27.759' W 81° 45.672'
Directions: From Barlow go west on OH-550 for 4.3 mi and turn right on CR-18. After 2.2 mi turn right on TR221 and in 1.7 mi make a sharp left on CR-18/Oak Hill Rd. Shortly make a right on TR570 and you will find the bridge in 1.3 miles.
Crosses: Wolf Creek, West Branch
Carries: TR570
Builder: Charles T. Shinn
Year Built: 1886 (R1998)
Truss Type: Multiple Kingpost and Burr Arch
Dimensions: 1 Span, 98 feet
Notes: Named after the local contractor, Charles T. Shinn, who built it, It looks great since the 1998 renovation. It has an all barn red exterior with a few well balanced windows.

World Index Number: OH/35-84-03
National Register of Historic Places: October 8, 1976

Goddard (Henry) Covered Bridge
Township: Fairfield
County: Washington

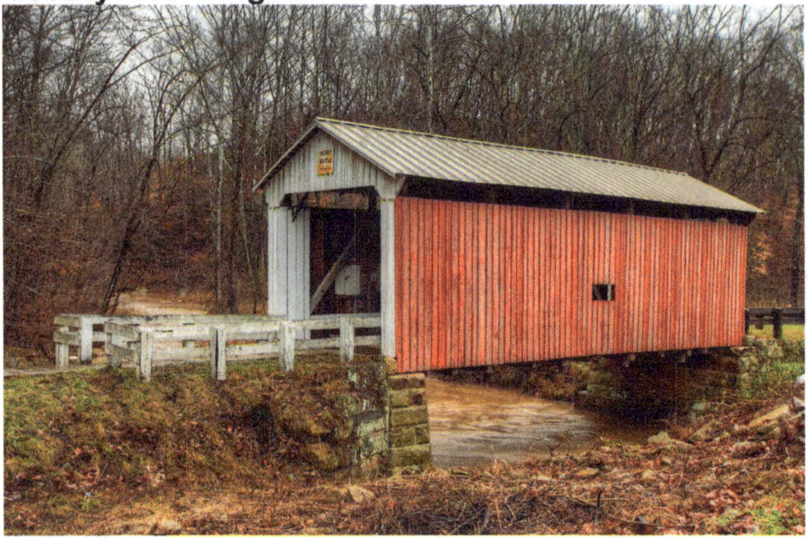

GPS Position: N 39° 23.005' W 81° 47.140'
Directions: From Bartlett go SW on OH-550 for 0.6 mi and turn right on TR77/Graham Rd. Proceed for 2.4 mi and turn right on TR-61 where you will see the bridge in 0.2 mi.
Crosses: Little Hocking River, West Branch
Carries: TR-61 (Bypassed)
Builder: E.B.Henderson
Year Built: 1894
Truss Type: Multiple Kingpost
Dimensions: 1 Span, 45 feet

Notes: This bridge was said to have been built after a small child drowned after falling from a rough footbridge. It is closed since being bypassed by a new structure.

World Index Number: OH/35-84-06
National Register of Historic Places: Not listed

Root Covered Bridge
Township: Decatur
County: Washington

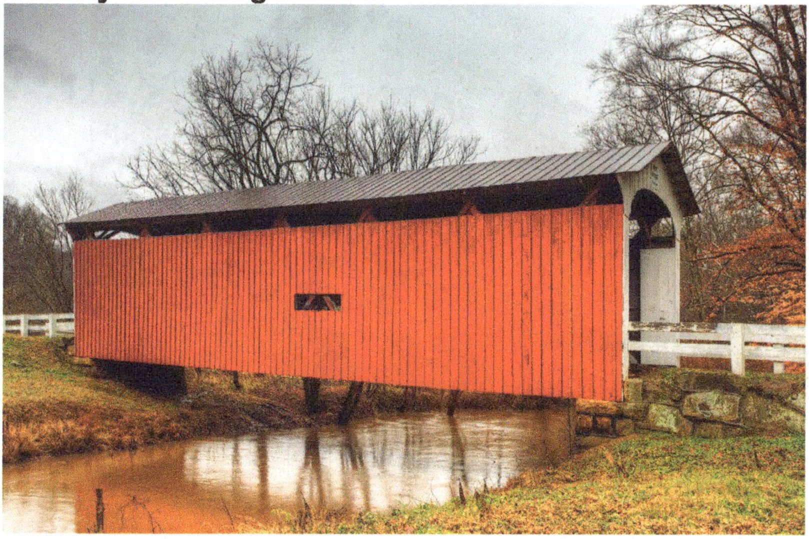

GPS Position: N 39° 20.521' W 81° 45.282'
Directions: From Cutler go south on OH-555 for 2.4 mi and turn left on CR-6/Burnett Rd and you will see the bridge
Crosses: Little Hocking River, West Branch
Carries: CR-6/Burnett Rd (Bypassed)
Builder: Rolla Merydith
Year Built: 1878 (R2001)
Truss Type: Multiple King
Dimensions: 1 Span, 65 feet

Notes: This bridge was named for a nearby community known as Root Town which has become a ghost town. It looks great after the recent renovation with barn red sides and white portals.

World Index Number: OH/35-84-08
National Register of Historic Places: March 27, 1975

Harra Covered Bridge
Township: Watertown
County: Washington

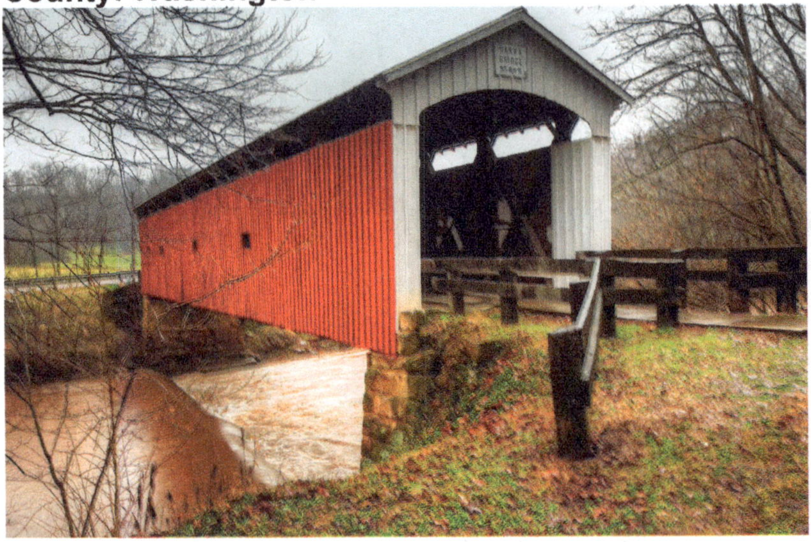

GPS Position: N 39° 29.281' W 81° 38.844'
Directions: From the town of Waterford go south on OH-339 for 4.1 miles and turn right onto Camp Hervida Rd/Township Rd 108 where you will find the bridge in 0.5 miles.
Crosses: Wolf Creek, South Branch
Carries: Camp Hervida Rd/Township Rd 108 (Bypassed)
Builder: Rolla Merydith
Year Built: 1878
Truss Type: Multiple King
Dimensions: 1 Span, 95 feet
Notes: The bridge was named after a nearby farmer, James Harrah whose property was the source for the stone used in the abutments by stonemason Billy Gamble. It was bypassed in 1981.

World Index Number: OH/35-84-11
National Register of Historic Places: October 8, 1976

Bell Covered Bridge
Township: Barlow
County: Washington

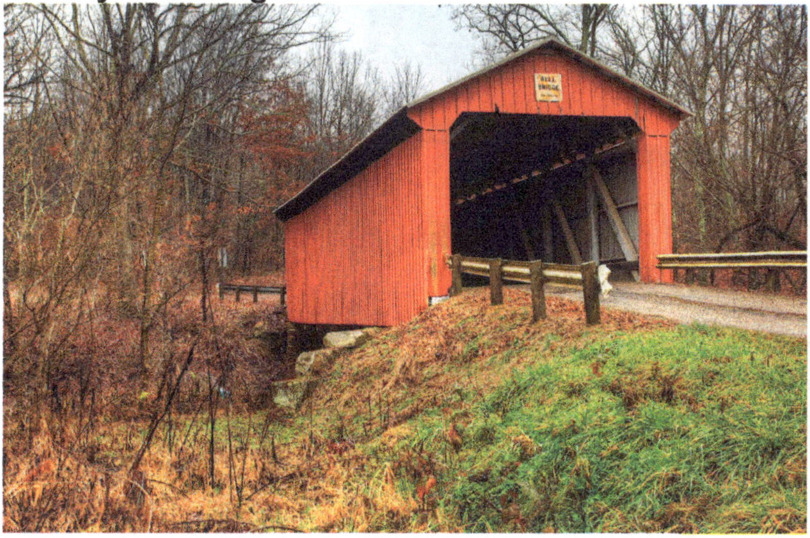

GPS Position: N 39° 25.887' W 81° 40.575'
Directions: From the town of Marietta go west on OH-550 for 10.6 miles and turn right on OH-339 and after 0.3 miles go left on Bell Rd. You will find the bridge in 2.8 miles.
Crosses: Wolf Creek, Southwest Branch
Carries: Bell Rd.
Builder: E.B. Henderson
Year Built: 1888 (R1998) (R2005)
Truss Type: Multiple Kingpost
Dimensions: 1 Span, 63 feet

Notes: The 1998 renovation has this bridge in great shape and it still carries traffic. It did have a temporary closing in 2005 when termite damage was found.

World Index Number: OH/35-84-12
National Register of Historic Places: Not listed

Mill Branch (Barlow) Covered Bridge

Township: Barlow
County: Washington

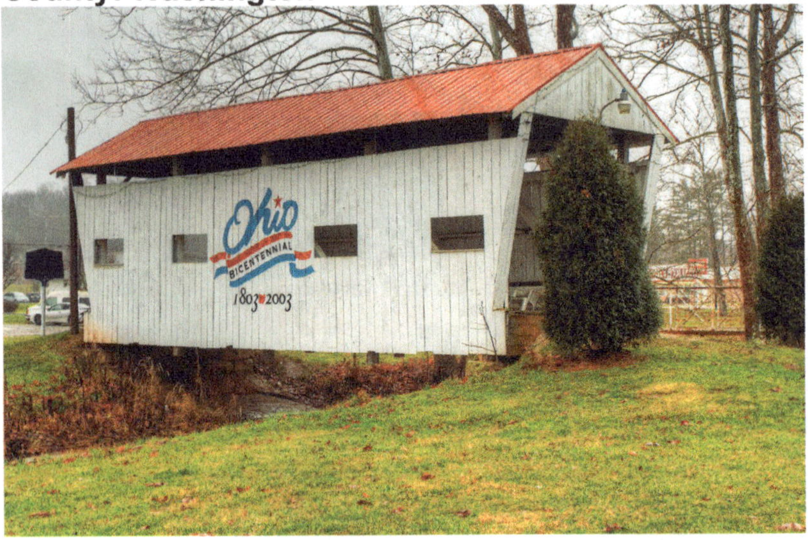

GPS Position: N 39° 23.987' W 81° 40.020'
Directions: In the town of Barlow, the bridge is located in the County Fairgrounds whose entrance is of OH-339.
Crosses: Unnamed Creek
Carries: Park Road
Builder: Not known
Year Built: 1885 (M1980) (R1999) (R2007)
Truss Type: Multiple Kingpost
Dimensions: 1 Span, 59 feet

Notes: The bridge was bypassed and abandoned in the 1960's and moved to its present location in 1980. During the Ohio Bicentennial, the logo was added to the sides. open to pedestrian traffic only.

World Index Number: OH/35-84-17
National Register of Historic Places: Not listed

Hills (Hildreth) Covered Bridge
Township: Newport
County: Washington

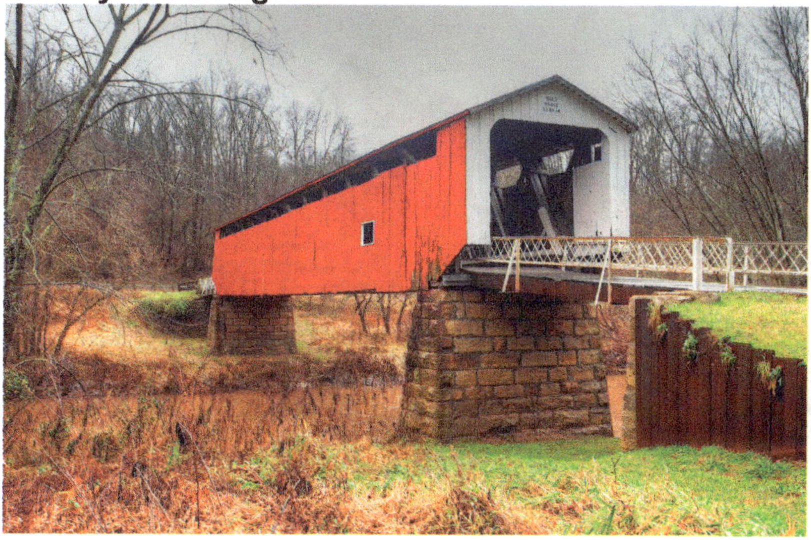

GPS Position: N 39° 25.644' W 81° 21.691'
Directions: From the town of Marietta go east on OH-26/Greene St for 3.9 miles and turn right on Hills Bridge Rd where the bridge is 0.3 miles.
Crosses: Little Muskingum River
Carries: Hills Bridge Rd (Bypassed)
Builder: August Borneman
Year Built: 1878
Truss Type: Howe
Dimensions: 1 Span, 112 feet

Notes: Standing on cut stone piers, it was bypassed in 1990. It is reported that the original construction was spaced out over 10 years. It stands quite high over the river below.

World Index Number: OH/35-84-24
National Register of Historic Places: February 8, 1978

Gallagher's Ford (Hune) Covered Bridge
Township: Lawrence
County: Washington

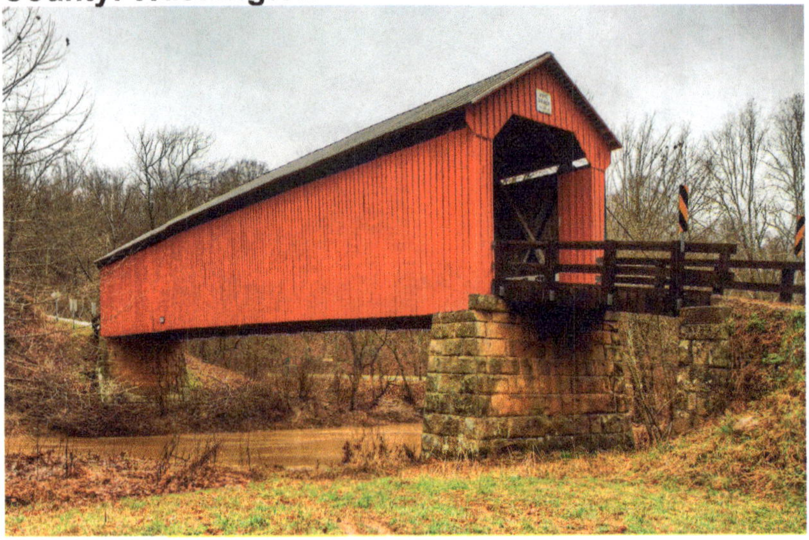

GPS Position: N 39° 30.657' W 81° 15.027'
Directions: From Marietta go northeast on OH-26 for 16 mi where you will see the bridge to the right on Smith Rd.

Crosses: Little Muskingum River
Carries: Smith Rd.

Builder: Rolla Merydith
Year Built: 1879 (R1998)
Truss Type: Unique
Dimensions: 1 Span, 128 feet

Notes: The structure sits high above the Little Muskingum River on cut stone piers with open approaches on each end.

World Index Number: OH/35-84-27
National Register of Historic Places: October 8, 1976

Rinard Covered Bridge

Township: Ludlow
County: Washington

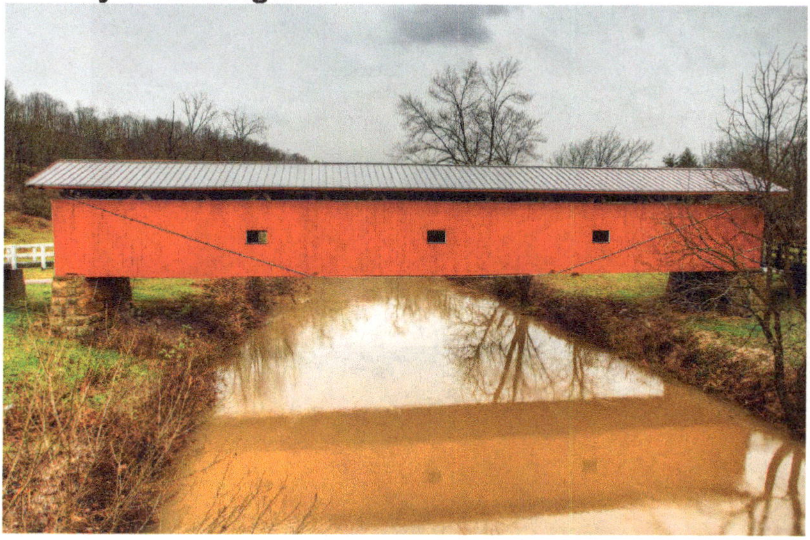

GPS Position: N 39° 32.208' W 81° 13.378'
Directions: From the town of Marietta go northeast on OH-26 for 21.4 miles to OH-406 where you will find the bridge on a bypassed section.
Crosses: Little Muskingum River
Carries: OH-406 (Bypassed)
Builder: Not known (Original Smith Bridge Company)
Year Built: 2006 (Original 1876)
Truss Type: Smith Double
Dimensions: 1 Span, 130 feet
Notes: This bridge has been hit by floods a few times during its history but two floods in 2004 destroyed it. The new bridge was completed in 2006 and it includes some of the original trusses.

World Index Number: OH/35-84-28#2
National Register of Historic Places: Not listed

Lockport Covered Bridge
Township: Brady
County: Williams

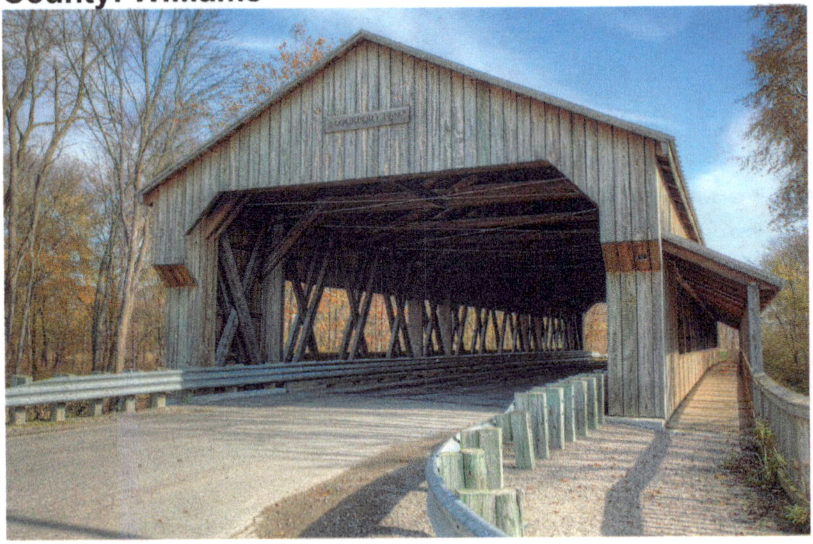

GPS Position: N 41° 32.788' W 84° 23.445'
Directions: Go north on CR-21 from OH-2 just north of Stryker for 2.2 miles and turn right on CR-25 where you will find the bridge.
Crosses: Tiffin River
Carries: CR-25
Builder: SSOE Group
Year Built: 1999
Truss Type: Howe
Dimensions: 3 Span, 176 feet
Notes: This bridge replaced an old steel structure and was constructed without the use of steel I-beams. It is a great looking structure and meets modern standards for height and weight requirements.

World Index Number: OH/35-86-06
National Register of Historic Places: Not listed

Montpelier Fairgrounds Covered Bridge

Township: Superior
County: Williams

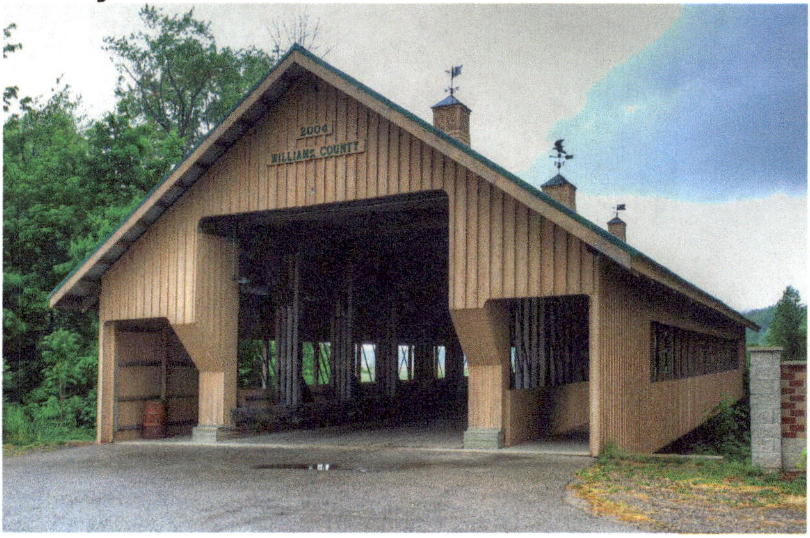

GPS Position: N 41° 35.436' W 84° 35.948'
Directions: Found in the town of Montpelier in the Williams County Fairgrounds off OH-107/main St. east of River St.
Crosses: St. Joseph River
Carries: Interior road

Builder: Mosser Construction, Inc.
Year Built: 2004
Truss Type: Pratt
Dimensions: 1 Span, 130 feet

Notes: The bridge collapsed during its original construction in December 2003 but was completed in 2004. It includes a pedestrian walkway at each side.

World Index Number: OH/35-86-26
National Register of Historic Places: Not listed

Parker Covered Bridge

Township: Crane
County: Wyandot

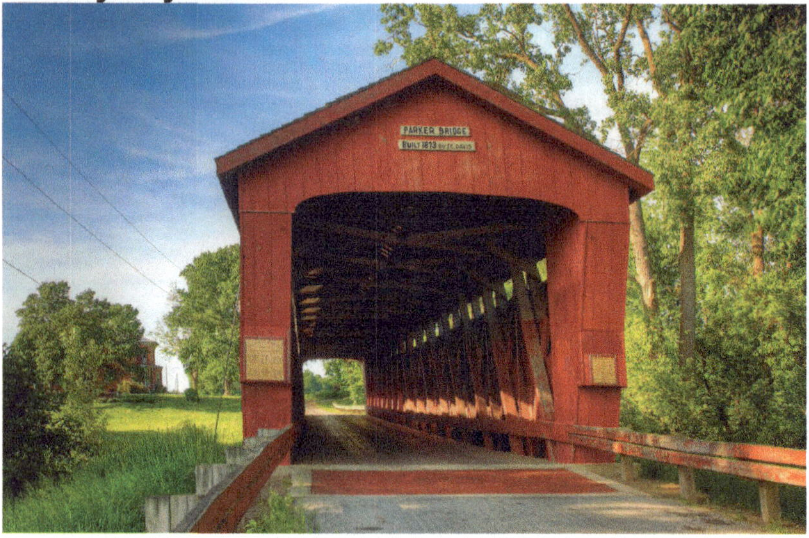

GPS Position: N 40° 54.150' W 83° 14.674'
Directions: From Sycamore, head SW on OH-67 W for 3.7 mi and turn right onto Co Hwy 39. After 1.2 mi, turn left onto Everhart Rd and the site is 0.2 mi
Crosses: Sandusky River
Carries: Everheart Road

Builder: Not known (Original J.C. Davis)
Year Built: 1992 (Original 1873)
Truss Type: Howe
Dimensions: 1 Span, 172 feet

Notes: This bridge was built to replace one from 1873 that burned in 1991. The interior and exterior is painted barn red.

World Index Number: OH/35-88-03#2
National Register of Historic Places: Not listed

Swartz Covered Bridge
Township: Antrim
County: Wyandot

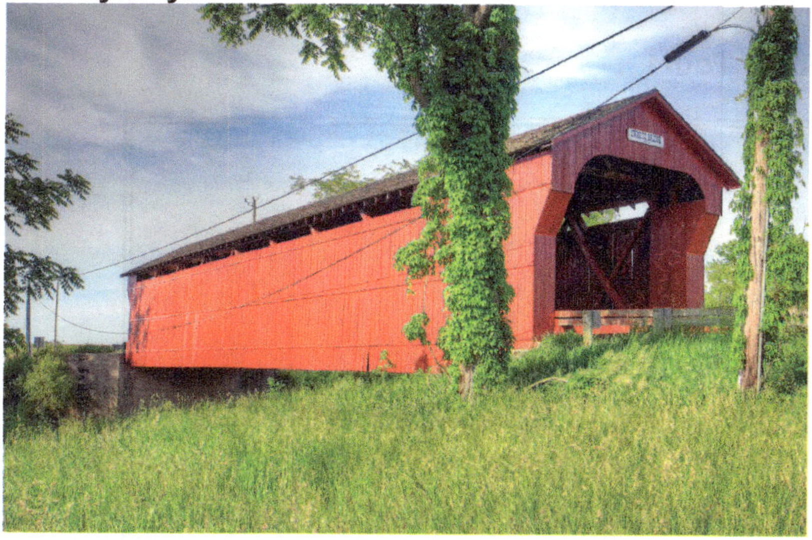

GPS Position: N 40° 46.272' W 83° 10.139'
Directions: From US-23, south of Upper Sandusky, take C-62 east for 3.8 miles and turn right on Co Hwy 130 where you will find the bridge in 0.3 miles.
Crosses: Sandusky River
Carries: Co Hwy 130
Builder: Moses Weymouth
Year Built: 1880 (R1994)
Truss Type: Howe
Dimensions: 1 Span, 93 feet

Notes: A great looking bridge in a peaceful country setting, it is barn red inside and out.

World Index Number: OH/35-88-05
National Register of Historic Places: October 8, 1976

Tours

The following tours provide an efficient manner of seeing multiple bridges

Ashtabula Tour1

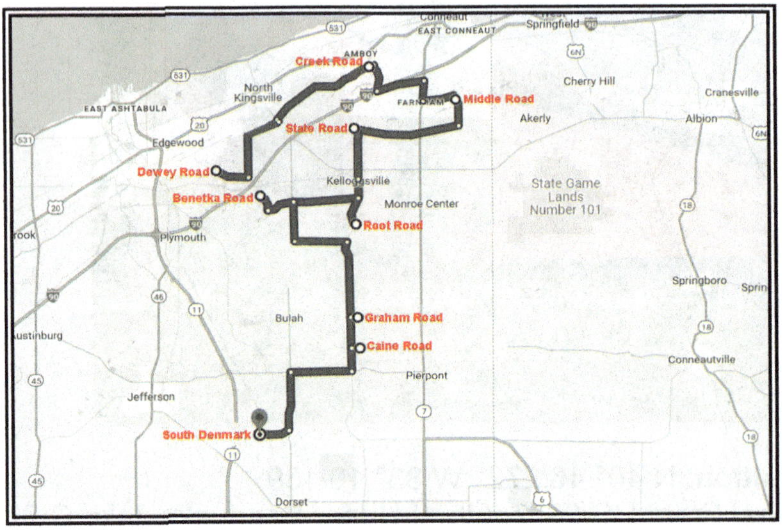

This tour visits 9 bridges with approximately 1 hours 15 minutes of driving. The last bridge on the tour is close to the start of Ashtabula Tour2

Dewey Road	N 41° 51.782'	W 80° 43.293'
Creek Road	N 41° 55.198'	W 80° 36.548'
Middle Road	N 41° 54.098'	W 80° 32.782'
State Road	N 41° 53.163'	W 80° 37.212'
Root Road	N 41° 50.001'	W 80° 37.169'
Benetka Road	N 41° 50.940'	W 80° 41.364'
Graham Road	N 41° 46.903'	W 80° 37.099'
Caine Road	N 41° 45.888'	W 80° 37.016'
South Denmark	N 41° 43.046'	W 80° 41.451'

Ashtabula Tour2

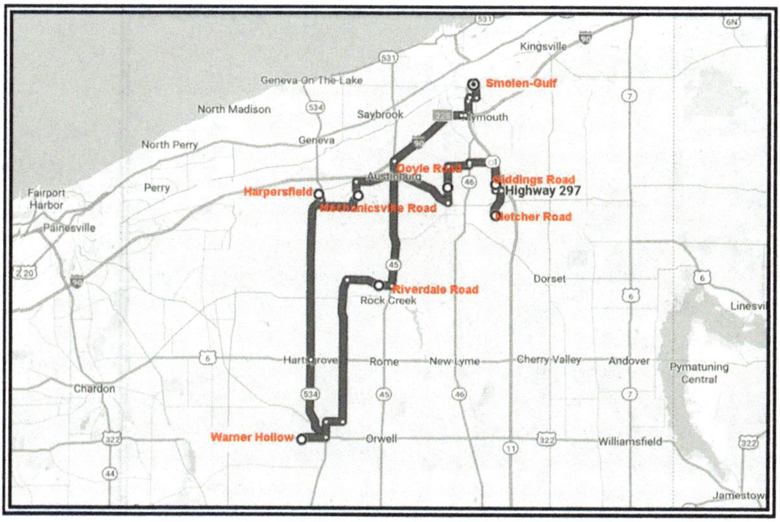

This tour visits 8 bridges with approximately 1 hours 45 minutes of driving. The first bridge on the tour is close to the last one of the Ashtabula Tour1

Netcher Road	N 41° 44.235' W 80° 43.935'
Giddings Road	N 41° 45.928' W 80° 43.951'
Doyle Road	N 41° 45.744' W 80° 47.423'
Mechanicsville Road	N 41° 45.295' W 80° 53.850'
Harpersfield	N 41° 45.330' W 80° 56.675'
Warner Hollow	N 41° 32.002' W 80° 57.820'
Riverdale Road	N 41° 40.397' W 80° 52.339'
Smolen-Gulf	N 41° 51.376' W 80° 45.574'

Fairfield County1

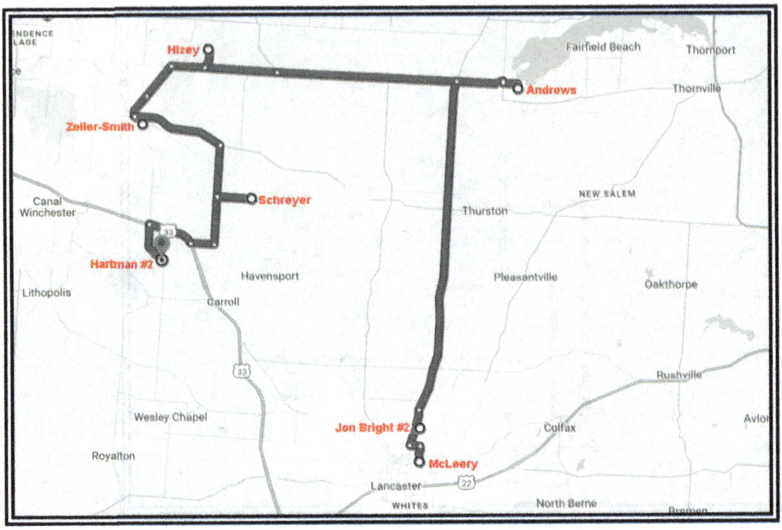

This tour visits 7 bridges with approximately 1 hours 15 minutes of driving. The last bridge on the tour is close to the start of Fairfield Tour2

McLeery	N 39° 43.498' W 82° 35.127'
Jon Bright #2	N 39° 44.361' W 82° 35.123'
Andrews	N 39° 53.789' W 82° 31.577'
Hizey	N 39° 54.866' W 82° 42.528'
Zeller-Smith	N 39° 52.814' W 82° 44.882'
Schreyer	N 39° 50.731' W 82° 40.686'
Hartman #2	N 39° 49.066' W 82° 44.216'

Fairfield County2

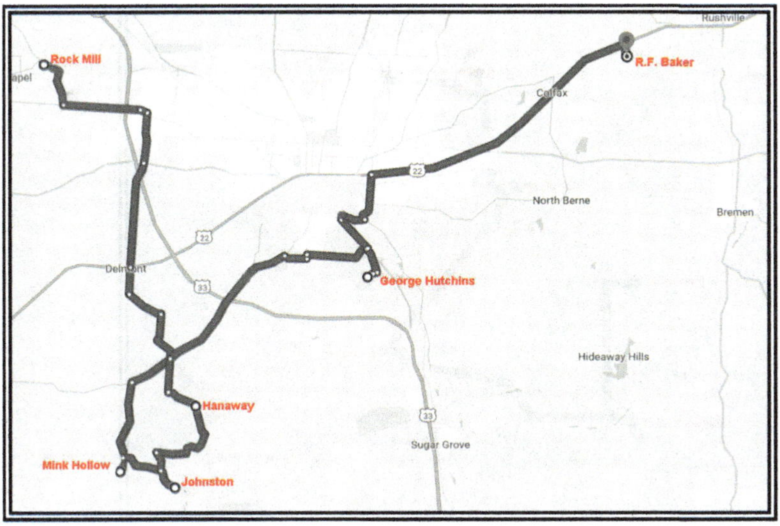

This tour visits 6 bridges with approximately 1 hours of driving The first bridge on the tour is close to the last one of the Fairfield Tour1

Rock Mill	N 39° 44.963'	W 82° 42.772'
Hanaway	N 39° 37.091'	W 82° 40.868'
Johnston	N 39° 36.778'	W 82° 39.511'
Mink Hollow	N 39° 38.350'	W 82° 39.002'
George Hutchins	N 39° 40.849'	W 82° 34.712'
R.F. Baker	N 39° 45.117'	W 82° 28.278'

Preble County Tour

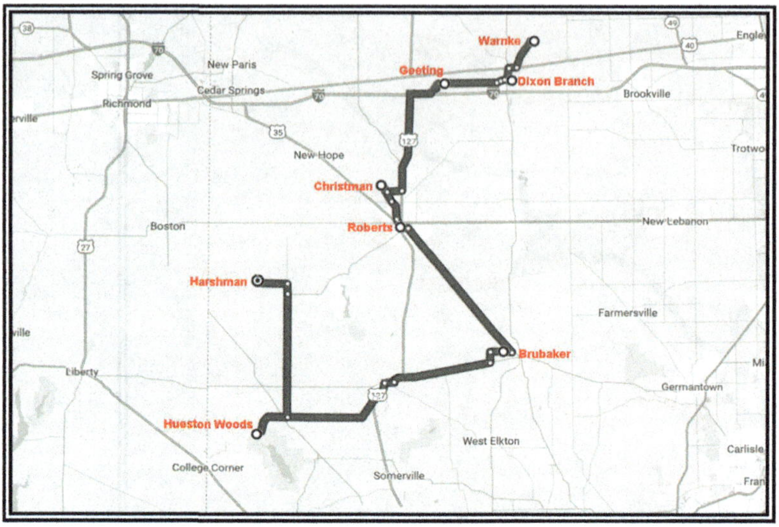

This tour visits 8 bridges with approximately 1 hours 30 minutes of driving.

Warnke	N 39° 52.420' W 84° 30.906'
Dixon Branch	N 39° 50.743' W 84° 32.135'
Geeting	N 39° 50.611' W 84° 35.842'
Christman	N 39° 46.236' W 84° 39.343'
Roberts	N 39° 44.428' W 84° 38.291'
Brubaker	N 39° 39.095' W 84° 32.610'
Hueston Woods	N 39° 35.515' W 84° 46.192'
Harshman	N 39° 42.129' W 84° 46.175'

Washington County Tour

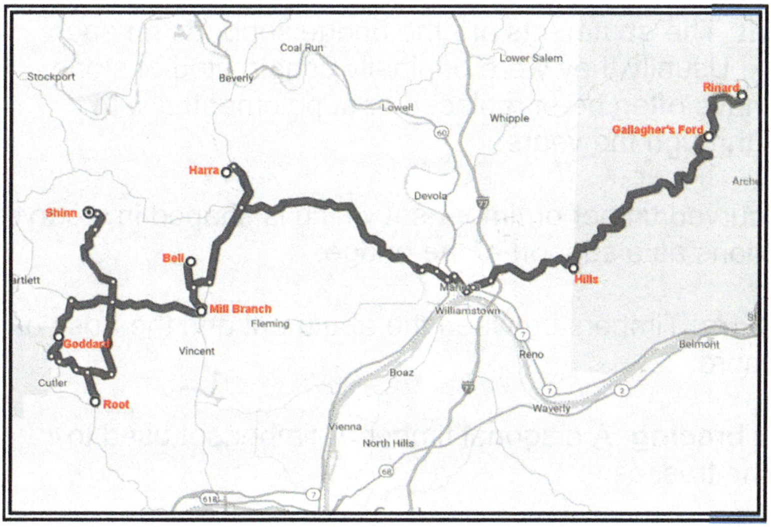

This tour visits 9 bridges with approximately 2 hours 15 minutes of driving.

Rinard	N 39° 32.208' W 81° 13.378'
Gallagher's Ford	N 39° 30.657' W 81° 15.027'
Hills	N 39° 25.644' W 81° 21.691'
Harra	N 39° 29.281' W 81° 38.844'
Bell	N 39° 25.887' W 81° 40.575'
Mill Branch	N 39° 23.987' W 81° 40.020'
Root	N 39° 20.521' W 81° 45.282'
Goddard	N 39° 23.005' W 81° 47.140'
Shinn	N 39° 27.759' W 81° 45.672'

Glossary

Abutment: The abutments are the bridge supports on each side bank. Usually they were originally constructed of stone but they have often been replaced or supplemented with concrete through the years.

Arch: A curved timber or timber set which is shaped in a curve and functions as a support of the bridge.

Bed timbers: Timbers between the abutment and the truss or bottom chord.

Brace or bracing: A diagonal timber or timber set used to support the trusses.

Bridge Deck: The roadway through the bridge.

Buttress: Wood or metal members on the exterior sides which connect the floor beams and the top of the truss. Used to keep the bridge structure from twisting under wind, water and snow loads.

Camber: A planned curve in the structure to compensate for the weight of the structure.

Chord: The horizontal members extending the length of the truss meant to carry the load to the abutments.

Dead load: The load of the weight of the bridge itself.

Deck: The pathway through the bridge used by pedestrians or vehicles.

Pier: Stone/concrete supports built in the stream bed to support the bridge

Portal: The bridge's entrances.

Post: The truss's vertical members.

Span: The bridge length measured between the abutments.

Treenails or trunnels: Pins or dowels turned from hardwood, driven into holes drilled into the members of the truss to hold them together. Also used in mortised joints.

Truss: The framework which carries the load of the bridge and distributes it to the abutments.

Truss Types

A Truss is a system of ties and struts which are connected to act like a single beam to distribute and carry a load. In covered bridges, these Trusses carry the load to stone abutments at each side and perhaps piers in between.

Brown

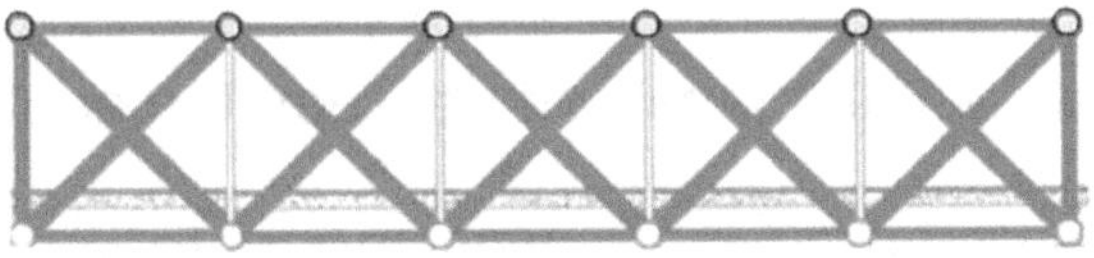

Brown

Josiah Brown Jr., of Buffalo, New York, patented this system in 1857.It consists of diagonal cross compression members connected to horizontal top and bottom stringers and is known for economic use of materials. It was only used in Michigan where there are a couple of surviving members.

Burr Arch

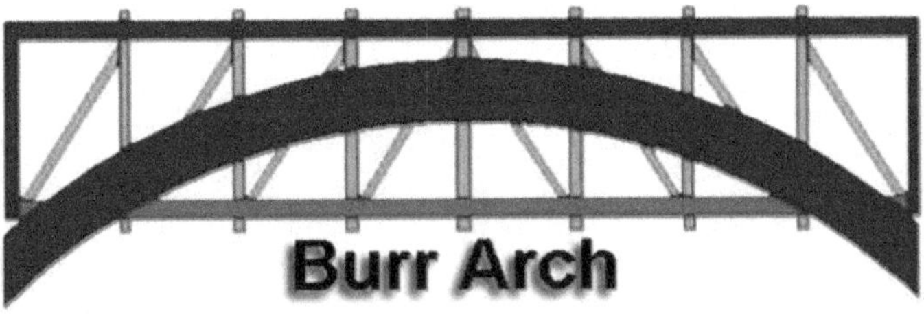

Invented in 1804 by Theodore Burr, the Burr Arch is one of the most commonly found structures in Covered Bridge design. It is often used in combination with multiple kingposts. The ends of the arch are buried in the abutments

Childs

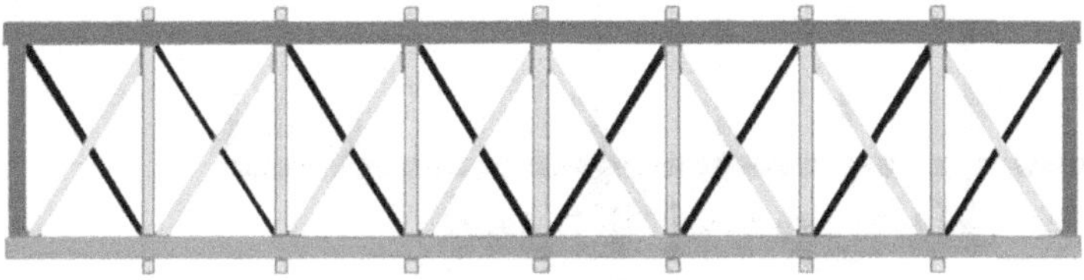

Childs

The Childs Truss System is essentially a multiple kingpost with half of the diagonal timbers replaced with iron bars.

Howe

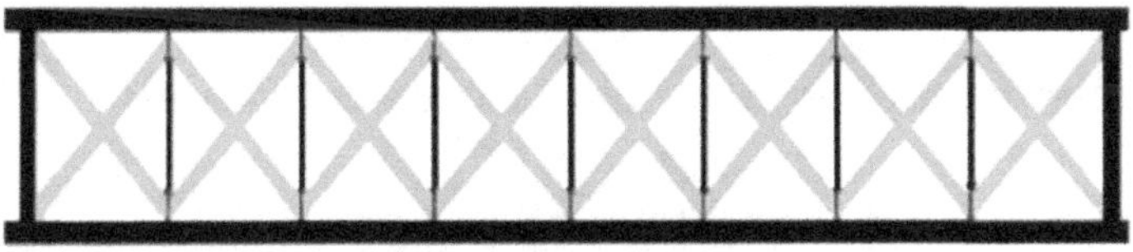

Howe Truss

The Howe Truss was patented in 1840 by William Howe. It involves the use of vertical metal rods between the joints of wooden diagonals.

Kingpost

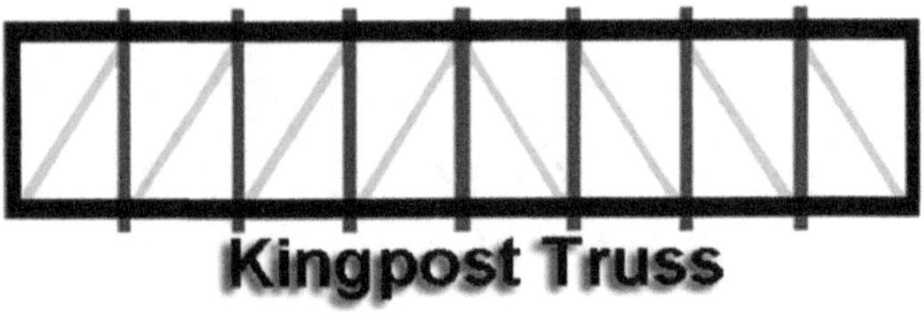

Kingpost is the simplest form of Truss with two diagonal members on a bottom chord, often with a vertical post connecting to the diagonals. The multiple Kingpost involves a series of Kingposts symmetrical from the bridges center. This allows for a much longer span.

Long

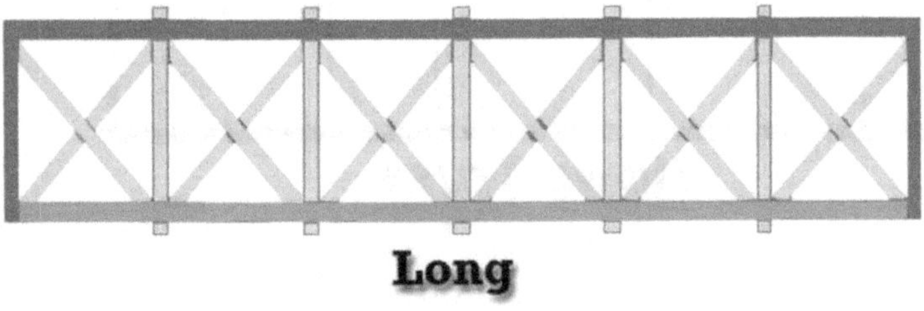

The Long Truss was patented by Stephen Long in 1830. It is a series of X shaped diagonals connected to vertical posts

Paddleford

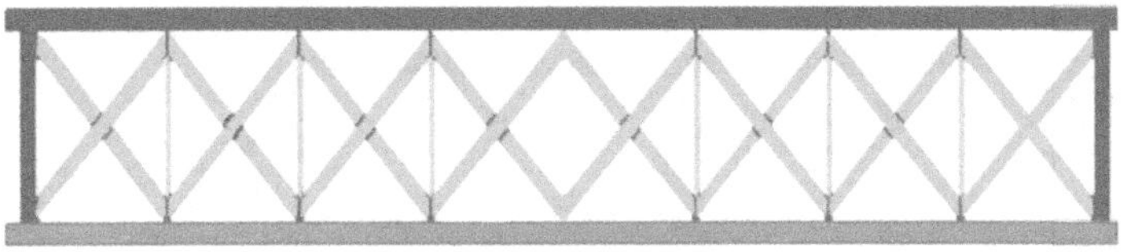

Paddleford

Peter Paddleford worked with the Long Truss system and eventually adapted it with a system of interlocking braces. he was never able to patent the system due to challenges from the owners of the Long Truss patent. However there are a number of New Hampshire and Vermont bridges which use the Paddleford system

Partridge

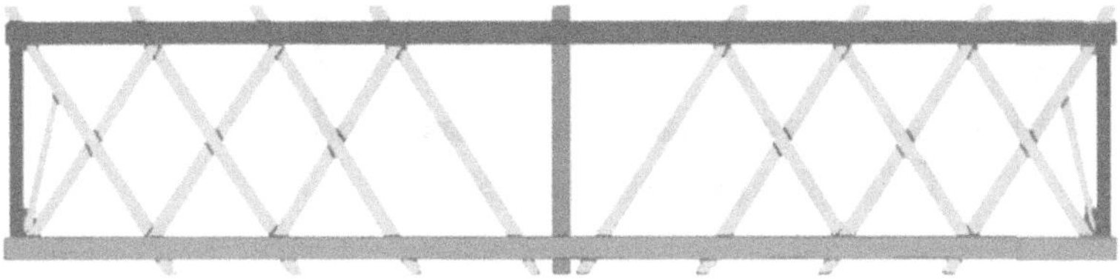

Partridge

Reuben L. Partridge received a patent for a design similar to the Smith system but adding terminal braces at the end and a central vertical member.

Pratt

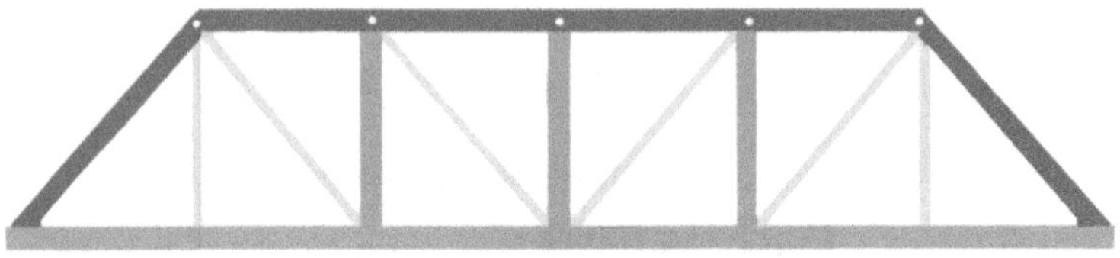

Pratt

The Pratt truss was patented in 1844 by Caleb Pratt and his son Thomas Willis Pratt. The design uses vertical members for compression and horizontal members to respond to tension.

Queenpost

Queenpost Truss

The Queenpost has the peak of the kingpost type replaced with a horizontal top chord which allows for a longer span.

.

Smith

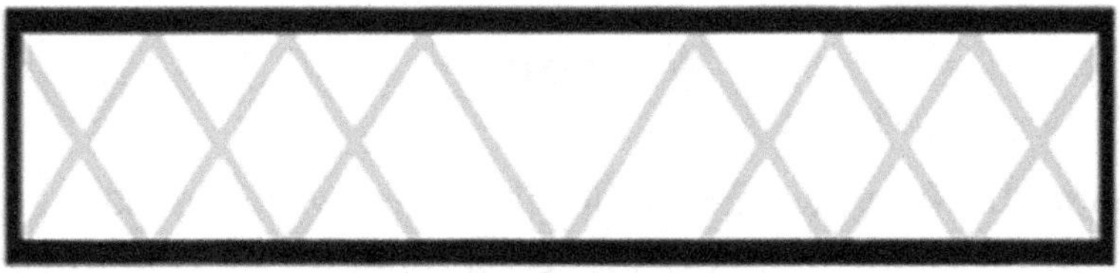

Smith Truss

Robert W. Smith received patents in 1867 and 1869 for variations of his system.

Town

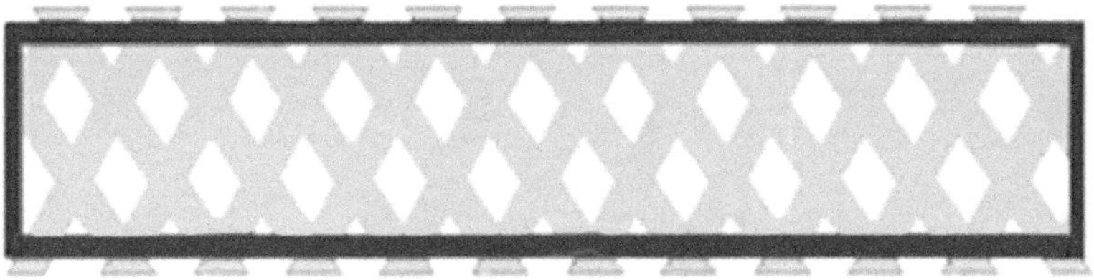

Town Truss

The Town or lattice system was patented by Ithiel Town in 1820. It involved a system of overlapping diagonals in a lattice pattern connected at the intersection by Tree nails or trunnels, wooden pegs or dowels. It had the advantages in that it could be constructed by unskilled labor and local materials could be used.

Warren

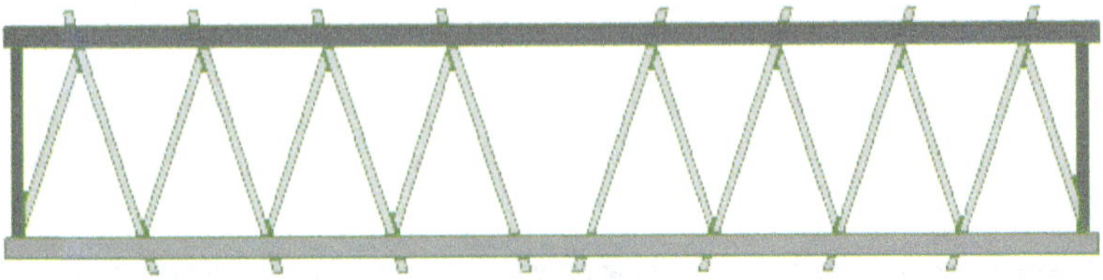

Patented in 1848 by two Englishmen, one of whom was named James Warren, it consists of parallel upper and lower chords with diagonal connecting members forming a series of equilateral triangles.

Recently Lost

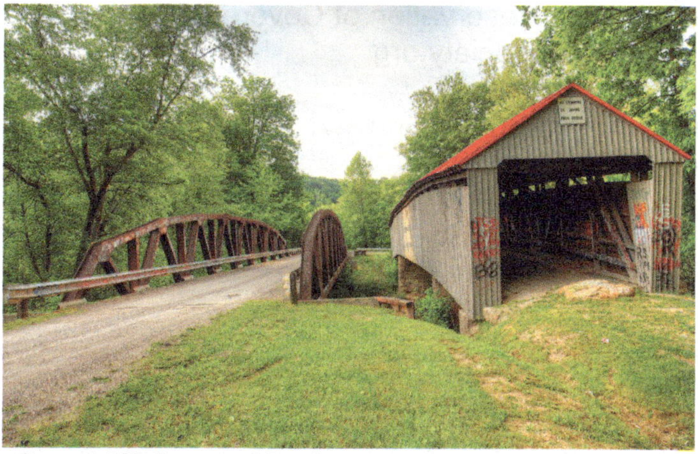

Geer Mill Covered Bridge

Geer Mill (Humpback, Penn's) Covered Bridge, Vinton County, OH/35-82-06x
Lost to arson June 6, 2013

Martin Hill (Scofield) Covered Bridge, Brown County, OH/35-08-21
Removed and placed in storage 1990

Roley School Covered Bridge, Fairfield County, OH/35-23-49x
Removed and placed in storage 1974

Shade Covered Bridge, Fairfield County, OH/35-23-20x
Collapsed March 6, 2022

References

National Society for the Preservation of Covered Bridges
http://www.coveredbridgesociety.org

New York State Covered Bridge Society
http://www.nycoveredbridges.org

Vermont Covered Bridge Society
http://www.vermontbridges.com/

Covered Bridge Society of Oregon
http://www.covered-bridges.org/

The Theodore Burr Covered Bridge Society of Pennsylvania
http://www.tbcbspa.com/

Indiana Covered Bridge Society
http://www.indianacrossings.org/

Ohio Historic Bridge Association
http://oldohiobridges.com/ohba/index.htm

Harold Stiver Image Gallery
https://haroldstiver.smugmug.com/Galleries/Themes/Covered-Bridges

Photo Credits:

Fatalfit, *Harshaville Covered Bridge*, **Nyttend**, *Manchester Covered Bridge*, **TC2U**, *Jon Raab Covered Bridge, Valentine Covered Bridge,* **T. Dillon**, *Jasper Road Covered Bridge*

All other images by the author

The Photographer's and Explorer's Series

Unless noted, there are Print and eBook editions available for the following.

Birding Guide to Orkney
Guide to Photographing Birds
Maine Lighthouses
Ontario Lighthouses
Ontario's Old Mills
Ontario Waterfalls

Alabama Covered Bridges (eBook)
California Covered Bridges (eBook)
Connecticut Covered Bridges (eBook)
Georgia Covered Bridges (eBook)
Indiana Covered Bridges
Maine Covered Bridges (eBook)
Massachusetts Covered Bridges (eBook)
Michigan Covered Bridges (eBook)
New Brunswick Covered Bridges
New England Covered Bridges
Covered Bridges of the Mid-Atlantic
Covered Bridges of the South
New Hampshire Covered Bridges
New York Covered Bridges
Ohio's Covered Bridges
The Covered Bridges of Kentucky (eBook)
The Covered Bridges of Kentucky and Tennessee
The Covered Bridges of Tennessee (eBook)
Vermont's Covered Bridges
The Covered Bridges of Virginia (eBook)
The Covered Bridges of Virginia and West Virginia
Washington Covered Bridges (eBook)
The Covered Bridges of West Virginia (eBook)

Index

Made in United States
North Haven, CT
10 November 2026

22088267R00108